Geography from the Air

First Published in 1953, *Geography from the Air* is the first book in English to explain and illustrate fully the use of air photographs in the study of geography. It sets out in simple form the limited amount of technical information which is necessary in order to enjoy fully the fascinating experience of training oneself to appreciate almost every aspect of the landscape shown on vertical air photographs. Thereafter the book consists of nearly 100 high quality air photographs, selected from almost every part of Britain, each of which is accompanied by a text describing in detail the features of general, as well as of geographical, interest which can be detected on the photographic prints.

The photographs have been grouped into sections detailing with many aspects of physical and human geography, such as the influence of geology on landforms, river erosion and river development, glacial action, coast forms, the relationships between relief and settlement, the appearance of varying agricultural and industrial regions and the shape and character of towns and villages etc. Apart from the explanation of this new geographical technique the selection of photographs is such that this book should serve as a valuable source of illustrative material for the study of the geography of Britain and the systematic treatment of physical and human geography.

Geography from the Air

F. Walker

Routledge
Taylor & Francis Group

First published in 1953
by Methuen & Co. Ltd.

This edition first published in 2024 by Routledge
4 Park Square, Milton Park, Abingdon, Oxon, OX14 4RN

and by Routledge
605 Third Avenue, New York, NY 10017

Routledge is an imprint of the Taylor & Francis Group, an informa business

© 1953 F. Walker

Publisher's Note
The publisher has gone to great lengths to ensure the quality of this reprint but points out that some imperfections in the original copies may be apparent.

Disclaimer
The publisher has made every effort to trace copyright holders and welcomes correspondence from those they have been unable to contact.

A Library of Congress record exists under LCCN: 53012391

ISBN: 978-1-032-86249-1 (hbk)
ISBN: 978-1-003-52201-0 (ebk)
ISBN: 978-1-032-86250-7 (pbk)

Book DOI 10.4324/9781003522010

GEOGRAPHY
FROM THE AIR

by

F. WALKER

Senior Lecturer in Geography in the
University of Bristol

LONDON: METHUEN & CO. LTD.
NEW YORK: E. P. DUTTON & CO. INC.

First published July 23rd 1953
Reprinted three times
Reprinted 1966

1.5
CATALOGUE NO. 02/5442/34 [METHUEN]
PRINTED IN GREAT BRITAIN
BY LATIMER TREND AND CO LTD, WHITSTABLE

CONTENTS

     A. River Erosion
          Stream Development
     B. Glacial Action

     The Relationships Between Relief, Soil Types and
         Settlements
     Valley Settlements
     Non-nucleated Settlements
     Settlement Patterns in a Scarp and Vale Landscape
     Evidence on Air Photographs of Former Settlement
         Patterns
     Settlement Shapes

     Economic Activity

ILLUSTRATIONS

All the photographs included in this book can be obtained from the Ministry of Housing and Local Government, Whitehall, London, S.W.1. All orders should quote a sortie number and the print number of the photograph

Chapter 1

THE GEOGRAPHICAL INTERPRETATION OF AIR PHOTOGRAPHS

The use of air photographs as an aid to geographical studies is a comparatively new technique and has only become common with the increased availability of suitable photographic cover since the end of the war. Even now, photographs of foreign countries are not generally or readily available and restrictions are frequently imposed on their use for security and other reasons. As far as Britain is concerned, however, very nearly complete photographic cover is available of this country and illustrative material is obtainable for a very wide field of geographical investigations. The photographs used in the present volume have therefore been selected entirely from Britain and have all, with the exceptions of Plates 4, 7 and 8, been supplied by the Air Ministry, who normally raise no objections to the use of such photographs for geographical purposes. The photographs included have, however, been specially selected because of their suitability as illustrations, from the point of view of scale, photographic quality, absence of cloud and haze, and the season and time of day at which they were taken, and it should not be assumed that such standards apply universally to the whole photographic cover of Britain. Nevertheless, the photographs selected are reasonably representative of the quality of prints available for a very large part of the country.

Normally, the ideal method of interpretation of air photographs is by means of the stereoscopic examination of adjacent pairs of overlapping photographs, and the use of single illustrative prints in the present volume is in no way intended to suggest that the study of individual prints is a substitute for the use of the stereoscope. Plates 32, 38 and 59 include pairs of photographs arranged for viewing through a simple hand stereoscope so that the advantages of this method of inspection may be appreciated. When adjacent pairs of air photographs are viewed through a stereoscope an image is seen which may be termed a space model of the landscape shown on the overlap of the two photographs. The viewer therefore enjoys the advantage of studying the equivalent of a perfect topographic model of the area in question, which includes every detail of the physical and human landscape and on which there is no distortion of the vertical and horizontal scales. The use of air photographs by such a method is therefore an extremely valuable partial alternative to the ideal of carrying out geographical studies in the field and by comparison with the latter it suffers only from two main limitations. In the first place it may take the student some time to become accustomed to the vertical view-point, and in the second place it is necessary to acquire some measure of skill in interpreting the meaning of the actual details visible, especially as the photographs are almost invariably in monochrome and it is necessary to learn the meaning of the gradations of tone on the photographs in terms of both the colour and surface texture of the objects photographed. (See below, p. 7.) At the outset, however, it is frequently the comparatively small scale of air photographs which seems to present major difficulties in interpreting the detail visible on the prints, in comparison with similar studies in the field. From Diagram 1 it will be seen that the scale of the air photograph is determined by the height from which the photographs are taken and by the focal length of the camera used.

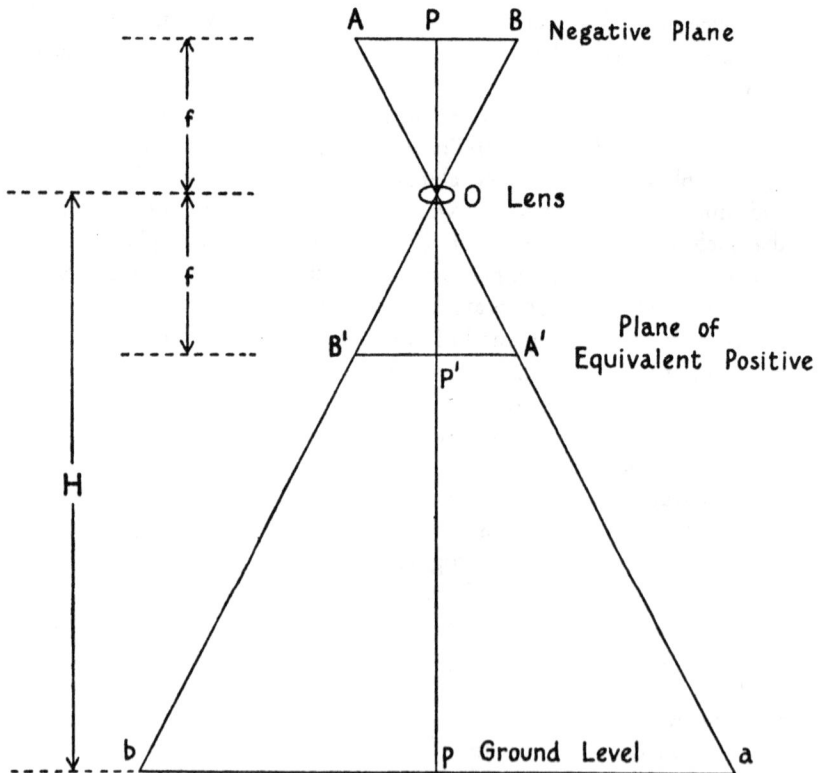

Diagram 1

The scale of the photograph is equal to $\dfrac{A'B'}{ab}$

and this, by similar triangles, $= \dfrac{OP'}{Op}$

$$= \frac{f}{H} = \frac{\text{focal length of the camera}}{\text{height of the camera above the ground}}.$$

Thus with a camera of focal length 12 inches flown at a height of 5,000 feet the scale of the print produced will be:

$$\frac{12}{5,000 \times 12} = \frac{1}{5,000}.$$

Plate 1 is an example of an air photograph obtained under very nearly the conditions specified, the focal length of the camera being 12 inches and the flying height approximately 4,600 feet—giving a general scale of 1 : 4,600. The photograph is included to demonstrate the large amount of detail which may be discerned on a print of this type, showing a moderately built-up urban area in the city of Bath.

The details of the lay-out of this portion of Bath which includes the Circus and the Crescent are immediately apparent and stereoscopic examination makes it possible to appreciate to an even greater extent than on the single plate much of the architectural character of the fine buildings in this part of the city. The photograph also serves to illustrate the displacement of the photographic image of objects of varying height away from the centre point of the print (see below, Appendix). As a result of this phenomenon, although the photograph is a very nearly vertical one, it is possible, towards the edges of the print, to see the displacement of the apparent position of the roofs of the buildings relative to their true plan position and in consequence the walls of the buildings actually appear, thereby adding somewhat to the possibility of studying even minor details of the appearance of their elevations. Two remarkable examples of this displacement are to be seen at A where the shape of a tower and small spire is visible, and near C where the frontages of most of the buildings can clearly be seen, especially in the case of the white building with its readily recognized darker windows.

At this scale details of vegetation are also clearly visible and the present photograph affords excellent examples of trees whose foliage is easily recognizable, especially in the case of the very fine specimens in the central garden within the Circus B. Moreover, the house gardens and in particular the allotments in the bottom part of the print are shown with remarkable clarity, and under the stereoscope such minor objects as small cold frames and rows of canes can be identified.

At various points on the photograph motor vehicles are to be seen, a number of them stationary in front of the buildings of the Crescent, in the streets near C and in the car park at D. By stereoscopic examination the type of vehicle can be recognized, and in one or two cases it is even possible to identify the distinctive design of certain makes of private car. Even the fine line markings on the lawn-tennis courts near D are visible on the original print for this plate, as are the white lines marked at various points on the roads throughout the area.

It will be readily appreciated that large-scale photographs of this type, when they cover areas of industrial and commercial importance make it possible to study the lay-out, function and traffic arrangements of individual industrial plants or transport undertakings with considerable precision, but it has been found impracticable to include a detailed photograph of such strategically important areas in Britain since its publication might constitute a danger to their security.

Naturally, photographs at a scale as large as that in Plate 1 have a very limited coverage, and the general cover of most of Britain is more normally at a scale of approximately 1:10,000, achieved by flying cameras of 12-inch focal length at a height of 10,000 feet, of 20-inch focal length at 16,650 feet, or of 36-inch focal length at 30,000 feet. All the remaining prints in the present volume are therefore based on photographs at an approximate scale of 1:10,000, but in a number of cases, particularly where prints have been made into a mosaic to cover larger geographical features, it has been necessary to reduce the scale by photographic means. In the case of Plate 2, however, the original scales have been maintained and photograph 2(a) is at a scale of 1:10,000, and photograph 2(b) at a scale of 1:9,840. The prints have been selected to show the amount of topographic detail which can be identified on good-quality prints, at normal scales, of typical rural country-side. Print 2(a) shows a small area (for map sheet number, co-ordinates and sortie, and print number, see page 16) on the Carboniferous Limestone of central Derbyshire, and the most striking feature is the remarkable clarity of the sharply defined stone-wall field boundaries characteristic of this area. At the points marked A, gates through the stone walls are made more prominent by the characteristic white 'flare' surrounding them—a mark created by the repeated trampling of the grass and earth round the gateways (see below under 'surface texture'). The details of farm and village buildings are visible and the patterns of walled gardens show up clearly in contrasting colour to the surrounding grassland. The railway tunnel is another feature common in the area and the dense shadows cast by the small viaduct at the entrance to the tunnel demonstrate one method by which it is possible to deduce the precise character of such structures appearing on air photographs. Print 2(*b*) shows an area in the Somerset levels (for location see page 16) and here too a vast amount of detail may be observed. Practically every line in the intricate pattern of drainage by which this area was brought under cultivation is visible, while at *A* the farm orchards, characteristic of much of Somerset, can be identified despite the date of photography, which was mid-December. At *C*, two large pylons carry power cables over the road and the long shadows of winter enable the lattice structure of the masts to be seen.

Although the relief of the land shown on air photographs can only be fully appreciated by means of stereoscopic inspection, nevertheless it is possible to identify many relief features on individual prints without the use of a stereoscope. Where such features are sharply defined, as in the case of cliffs or steep-sided valleys, the precise relief form may often be ascertained from the length and shape of shadows, and this applies equally to such topographical features as trees and buildings (as for example the viaduct on Plate 2(*a*) below). Provided the date and time of photography is known it is possible to calculate the altitude of the sun at the time of exposure so that by precise measurement of the shadow length, using a magnifying optical measure reading to 1/10th mm., the exact height of the object in question may be calculated. Obviously, this method has only limited application in the study of relief features themselves, but it has specialized applications of very great value. It has been used, for example, in the measurement of tree dimensions for the estimation of stands of timber, it can be applied to the measurement of rock faces in mountainous country where the depth of shadow itself makes other methods difficult, and it has a very wide range of use for military purposes.

Where surface morphology is composed of gentler relief features no sharply defined shadows are formed, but the photographs still show clearly the modelling of the terrain. This effect is produced by the soft gradation of less-defined shadows in a way comparable to that which is supposed to exist when hill shading is incorporated in certain types of relief map, where a source of light is presumed to cast slight shadows on the relief forms. Plate 4 is designed to show an area of comparatively gentle relief forms where the air photograph gives a nearly perfect impression of the landscape character (for location details see p. 16). Isolated crags and rock outcrops are accentuated by the creation of deep shadows, but the rounded hills of the area are equally defined as are the valleys and stream courses.

In order to appreciate the value of the information to be derived from such a photograph in the study of the physical landscape it should be compared with the information of a similar kind to be derived from the best available map, and to this end Plate 5 consists of a map of the same area at precisely the same scale as the photograph. The map, which is enlarged from the Ordnance Survey map at a scale of 1:63,360, includes all the relief information shown on that map and the comparison of the two plates demonstrates the inadequacy of its contour interval for the appreciation of the details of relief which are evident on the photograph. The map, despite its great accuracy, gives the impression of parallel valleys of regular and smooth cross-section, and the only evidence of the essentially hummocky and irregular relief is in the form of one or two isolated small contour 'rings', whilst the slight incision of the stream courses is obviously not capable of being represented by means of such a contour interval. Both these features are, however, of great importance, not only from the point of view of appreciation of landscape type and land use, but also in the study of the physical history of an area[1] which has been subjected to heavy glaciation and whose streams have undergone a number of changes of effective base level. (See below under Plate 38 which also concerns this area.)

The degree of surface modelling and therefore the identification of the smallest changes of height and slope are clearly conditioned by the intensity of light and shade and therefore by the date and time of photography. In Plate 4 the exposure was made during the afternoon in late April with resulting shadows of moderate length and intensity, but an example of an extreme case is to be seen in Plate 92(b) where the very long and deep shadows cause even field furrows to stand out sharply.

From a geographical point of view the appreciation of colour variations in landscape is almost as important as the recognition of relief forms, particularly when it is applied to the interpretation of soil colour, natural vegetation and cultivation patterns. Unfortunately colour film has so far been seldom applied to air photography and the number of coloured air photographs available is so small that at present the method is not generally available for study. The universal use of some form of panchromatic film and appropriate filters for air photographs, however, makes the identification of most natural colours as simple as on a good ground photograph, though only prolonged experience will make it possible to differentiate between the colours of various growing crops

[1] The rock in the district covered by the photograph is dolerite, which in this area tends to produce irregular relief.

for example. In many cases, however, colour contrasts are sharp and clear, and this is particularly true in the case of trees with different colours of foliage. Plate 3 shows a small area of afforestation in Southern Scotland (see p. 16) where the zones of different trees stand out as clear patterns of colour and where vegetation patches on the near-by rocky surface of Screel Hill are largely distinguishable because of their colour against the bare areas. A further point of interest in Plate 3 is that it demonstrates the location of the forestry areas on the more sheltered Eastern slopes of Screel Hill, a common practice in such schemes on the Southern flanks of the Southern Uplands. It is of interest to note, too, that the comparative ease of distinguishing tree types on air photographs makes this an ideal method of preliminary reconnaissance of existing stands of timber, particularly in areas of dense or inaccessible forest where the definite identification on air photographs of isolated patches of economically valuable timber can precede the planning and construction of the very costly roads or tracks along which it is to be extracted.

Air photographs enjoy a special advantage over maps or ground photographs in that the texture of the surface of any object photographed may lead to a sharp colour distinction on the print. Diagrams 2 and 3 show the cause of this phenomenon.

In Diagram 2 the greater part of the sunlight reflected from a smooth surface passes along a path such that the angle of reflection is equal to the angle of incidence, and therefore with the camera in position (a) such an object appears white or a very light tone of grey, whereas with the camera in position (b), or any other such position, the object appears black, or very nearly so. In Diagram 3, on the other hand, the scattering of the light rays caused by the irregular surface results in objects with such a surface appearing some shade of grey whether the camera is in position (c), (d), or any other position. The exact depth of this grey is determined by the degree of irregularity of the surface in question and by the true nature of this irregularity, a good example being the different shades of grey appearing on grass fields where varying lengths and types of grass produce a surface of varying irregularity. An important consequence of this characteristic of air photographs is that any activity which tends to modify natural surface textures will be recorded on the photographic plate. Thus even an individual walking across a grass field tends to flatten an appreciable number of blades of grass which then present a broader and smoother surface to the light and occasion the appearance of a white streak on the photographic print, a fact of some importance in the use of air photographs from a military point of view. The white 'flare' marks round field gates on Plate 2 (a) (below) are therefore explicable

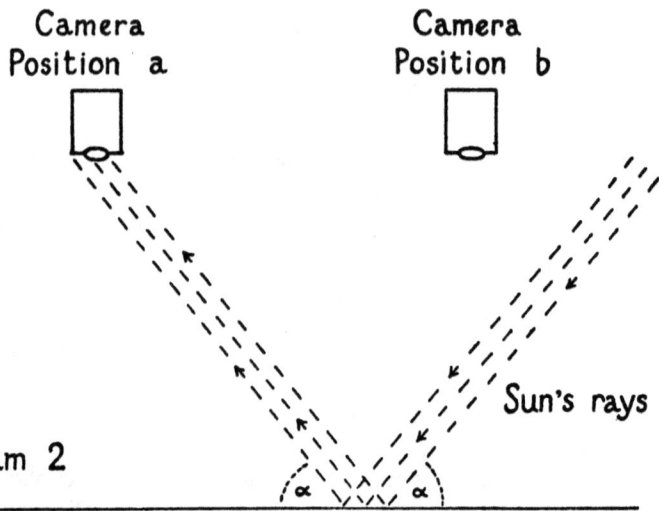

Camera
Position a

Camera
Position b

Sun's rays

Diagram 2

α   α

Smooth Surface (Still water, centre of metalled road etc.)

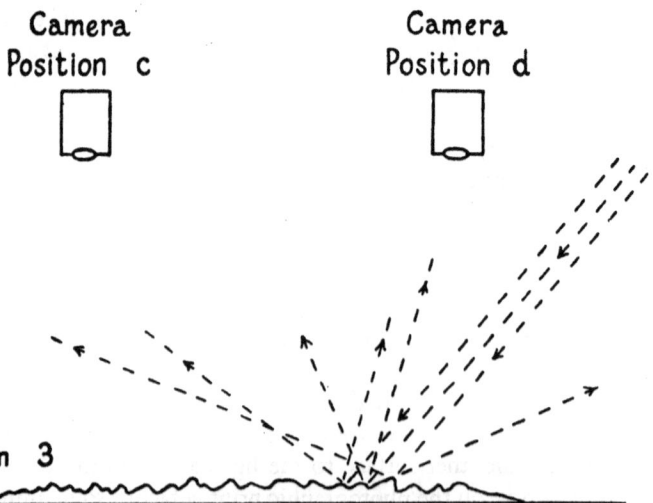

Camera
Position c

Camera
Position d

Diagram 3

Irregular Surface (Rippled water, grass, rough ground etc.)

in that they represent the areas round the gate where grass and soil have been trampled smooth. Variation in surface texture may, of course, be occasioned by a large number of causes, many of them of great interest to geographers and archaeologists and others. Variations in soil type, in water level and in the character of newly turned earth, all have consequences which produce slight differences of grey tone and make it possible in most cases to identify their boundaries, and examples are included later where it is possible to do this. From an archaeological point of view surface morphology and shadows are often the clearest indication of sites, but the existence of unsuspected earthworks beneath apparently smooth surfaces have frequently been discovered on air photographs because the former disturbance of the native soil has produced minor variations in vegetation which are almost impossible to detect even in the field but which leave a tell-tale tonal variation on the photograph, outlining the frequently characteristic shape of the archaeological site in question. Plate 6 demonstrates the effect of variations of surface texture on golf-links in the Braid Hills near Edinburgh. The light zones marked A, B, C, D, E are areas on greens and the fairway where the grass has been mown short and where the scattering effect on the light is at a minimum, the darker zones representing the areas where the grass is longer. The Braid Hills and Blackford Hill (in the top right-hand corner) are on parts of the igneous rocks of the Edinburgh region, and it is of interest to note that the lavas of the former area provide a naturally hummocky surface forming an ideal setting for the golf-links, and it may be noted that only where the links leave the area of the lava at D and E has it been thought necessary to introduce artificial hazards in the form of bunkers.

In the study of human geography the main advantage of the use of air photographs lies in the fact that the photograph inevitably records all visible evidence of human activity in relationship to its physical setting. In the case of maps or diagrams, on the other hand, it is almost always necessary not only to exercise selection of the topographical detail to be shown but also to exclude from normal topographical maps vital distributions such as variations of rock or soil character, the type of natural and cultivated vegetation, and many of the minor relief forms which may be fundamental to an understanding of certain forms of human activity, all of which can normally be deduced from good air photographs. On the other hand it must be emphasized that, just as the study of physical geography from air photographs is not a substitute for detailed field study, so the interpretation of the photographs from the point of view of human geography demands an adequate knowledge of the historical geography, space relations and present-day human geography of the region in which the photograph was taken if totally unwarrantable conclusions are not to be reached. In other words, the best use of air photographs in this connection may well be to exemplify established interpretations of the human geography of an area in its relationship to the physical environment rather than to attempt new interpretations based solely on the photographic evidence.

The geographical interpretation of air photographs, particularly from the point of view of physical geography, may be greatly enhanced if detailed and precise measurements (of minor relief forms for example) are made from the photographs. The science of photogrammetry has as its principal objective the carrying out of this precise survey work and measurement from air photographs, and in the Appendix[1] a brief description is given of the equipment used in such work (three examples being illustrated in Plates 7 and 8). It is perhaps sufficient to state at this point that provided suitable photographic cover is available for use with the appropriate photogrammetric equipment the most precise measurements required for geographical purposes may be made as a matter of normal routine.

[1] See below, pp. 105-6.

Chapter 2

GEOLOGICAL INFORMATION ON AIR
PHOTOGRAPHS

The application of air photographs to the study of geology is becoming a highly specialized subject and one which cannot properly be discussed fully within the scope of the present work. Provided that a certain amount of identification of rock types is carried out on the ground it is in many cases possible to carry on the further processes of geological mapping by means of air survey. This method is clearly of maximum use where the outcrops and structures to be mapped are not masked by drift or vegetation or by thick soil development. So far, air survey for such purposes has been made use of mainly by economic geologists and especially those concerned with the mapping of oil-field areas, and it has been fortunate that in the oil-producing regions of the Middle East, for example, geological and topographical conditions have resulted in the air photographs showing the details of the structures at the surface with startling clarity. In this case, of course, clear, dry air and prolonged good flying weather have also made for the acquisition of sharply defined, high-quality photographs. In Britain suitable areas for this type of work are less common, but in parts of Northern and Western Scotland and in the Western Isles, for example, there are many instances where air survey methods for geological mapping may well be adopted, and Plates 9(*b*), 12, 30, 57 and 58 may be taken as indicative of the appearance of such districts on air photographs.

For the geographer, however, the identification of geological phenomena on air photographs can be of great importance where the effect of geology on landscape or on human activities is considerable and where this fact can be demonstrated to best advantage from the air. In most cases the geological facts which are evident on the air photograph can be equally well observed on the ground, and the sole advantage of air photography lies in the fact that the unusual view-point makes it easier to appreciate the geographical significance of the features portrayed. On the other hand there are many instances where air photographs show geological distributions which are difficult to trace on the ground but which are of great importance to the geographer.

The two photographs on Plate 9 are intended to demonstrate cases of sharply defined geological boundaries where the effects of the junction produce easily recognized consequences in the landscape and land utilization of the area in question. Photograph (a) shows a small portion of the Northern edge of the Granite mass of Dartmoor near the village of Belstone (see p. 16) and it will be seen, particularly in the top right-hand corner of the print, that the marked break of slope associated with the edge of the Granite produces an easily recognizable feature on the photograph. Farther East the junction is not so clearly seen in the relief of the area but the remarkable contrast in land utilization between the moorlands and rough pasture of Dartmoor and the enclosed fields of grass and arable on the plain makes the significance of the geological boundary equally evident. The small valley running Northwards from Dartmoor provides a sheltered area where settlements have developed on the thicker valley soils overlying the Granite, but, in the main, settlement in this area is confined to the village of Belstone on the plain where a number of minor roads, skirting the Northern edge of Dartmoor, converge.

Photograph (b) demonstrates a similar geological junction where the vertical view serves to emphasize a relief feature which is remarkable for its abruptness even when seen on the ground or on a map. It is a portion of the Southern edge of the Ochil Hills immediately East of the village of Menstrie and marks the junction of the contemporaneous igneous material of the Ochil Hills with the alluvium of the plain. In this case the sharpness of the contrast is accentuated by two other circumstances, firstly the fact that the Southern edge of the igneous material is marked in this area by a continuous line of faulting, and secondly the extreme flatness of the plain to the South, a fact which can be appreciated from the remarkable straightness of the road and railway at the bottom of the photograph and the absence of cuttings or embankments on either. This flat alluvial plain forms part of the so-called 50-foot raised beach which is very extensive in this part of the central lowlands of Scotland and therefore the edge of the hill mass coincides very closely indeed with the former coastline—a phase of the area's physical history which can be readily visualized on the photograph. The excavation of deep gully-like valleys by the streams flowing Southwards from the Ochil Hills has led to the development of well-marked alluvial cones where they emerge on to the plain, and two of these are evident on the photograph at A and B. In the latter case the cone is virtually outlined by the straight rows of trees, and it is of interest that a site on the slightly higher and drier soils of the cone has been chosen for Balquharn House. A study of the topographical and geological maps of the area will reveal that similar alluvial cones at the exits of other valleys on the Southern side of the Ochil Hills form the sites of a line of settlements and towns along the junction of the hills and the plain, a much larger cone immediately to the West of the area shown on the photograph forming the site of the village of Menstrie for example. It is also evident, in part, from the photograph that whereas the railway and present main road have been planned as straight lines across the alluvium, the older, curving road, partly hidden by trees, appears to have developed as a link between the settlements on the alluvial cones. The photograph also shows up the way in which tree growth has been largely confined to a narrow belt on the lowest slopes of the hills and above the predominantly arable land of the flat area of the alluvium.

For reasons explained above (see pp. 6–9) the identification of colour and texture differences on air photographs makes it possible, in some cases, to trace geological boundaries where there is relatively little change of relief along the line of demarcation, and in contrast with Plate 9 the photograph in Plate 10 shows a virtually flat area where geological divisions can be distinguished largely because of variations in vegetation and land use. The area shown is part of the low-lying alluvium round Wigtown Bay in South Scotland, and the two outlying areas of peat in Carsegown Moss and Borrow Moss are outlined at A and A_1. The alluvial areas which cover almost all the remainder of the photograph are now in high cultivation, mainly as arable land, and the sharp colour contrast between fields recently under the plough, which appear white and those with a grass ley which appear grey, serves to demonstrate the effects of their different colour and surface texture. The fields themselves are large and, except where bounded by the streams meandering over the alluvium, straight-sided, a feature common in such areas of formerly ill-drained land which have been brought into cultivation comparatively recently, and it may be noticed that there is evidence on the photograph at E of the nature of the land in this area before it was cultivated since the shadowy marks at this point represent traces of former water channels, none of which is now evident to the observer on the ground. The ease of distinguishing the peat areas arises mainly from the fact that they are still largely uncultivated (an example of cultivated peat land is included later, see Plate 51) but attempts to improve them can be traced in the pattern of drainage cuts round their edges and in the longer straight drainage channels cut across them. The former liability to floods through-out this area resulted in the road D keeping to the edge of the slightly higher ground, where it may be noticed, in the field which appears white on the photograph, isolated patches of rock project through the alluvium and appear as small dark patches. By the time the single-track railway C was constructed, most of the drainage was completed and it was possible to build the line along the lower ground, though speeds often have to be restricted on this track, especially in the parts of the line where it proved impossible to avoid peat areas completely.

In addition to showing the effects of geological junctions of both major and minor importance, air photographs may, under favourable circumstances, be used to study the effects of actual structures within rock types. Provided that good drift-free photographs are available, it is possible, for example, to measure by photogrammetric methods the dip of exposed rocks and to plot their strike as precisely as would be required for large-scale geological mapping. To the geographer, however, it is the relationship between these structures and surface morphology and their effect upon topography and human activity that is primarily of interest and these features are commonly apparent on air photographs even where complete rock exposures are not evident at the surface. For example on Plate 17 the general dip of the Carboniferous Limestone in the Northern Mendips near Shipham may be seen at the side of the valley immediately above the large quarry and the relationship between this dip and the morphology of the Northward-facing scarp of Mendip appreciated, whilst the siting and working lay-out of the quarry itself can be understood from the photograph in the light of the evident structure of the immediate area.

Plate 11 shows a portion of the scarp of the South Cotswolds at the village of Horton and has been selected to demonstrate the way in which the erosion of the very gently dipping beds of the Middle and Upper Lias and of the Oolites has produced a broad zone of wide steps rather than a sharply defined single scarp face. The Oolites dip gently in the general direction of the arrow, but only the Inferior Oolite appears on this photograph, the Great Oolite appearing a little way beyond the right-hand edge of the print, but producing a negligible break of slope which can only be discerned with great difficulty both on the ground and on the succeeding photograph of this run. The scarp face of the Inferior Oolite is indicated by the right-hand dotted line on the print and it will be seen that it forms an appreciable break of slope. The lowest relief feature in this portion of the Cotswold edge is the minor scarp face indicated by the second dotted line which forms the edge of the Middle Lias Marlstone ledge, below which the Lower Lias Clays appear at the left-hand side or the photograph. Between the Middle Lias Marlstone and the Inferior Oolite there occurs the sandy material corresponding stratigraphically to the Midford Sands which in this part of Cotswold rarely forms a recognizable topographical feature, though in this photograph at B a slightly higher piece of ground on which is situated the upper part of Horton village can just be distinguished and marks a tongue of the Sands lying on top of the Marlstone ledge.

Apart from its effect on the relief of the area the succession of gently dipping rocks eroded into their present form has had a profound effect on the human geography of the district, since it was on the Marlstone ledge that most of the early settlement developed (see below, pp. 82-3) including the original settlement of Horton which is now the site of Horton Court at the top right-hand corner of the print, whilst the distribution of lynchets at C, C_1 and C_2 indicates the importance of the slope at the junction of the Marlstone and the Inferior Oolite in the early agricultural development of this area. The field marked A lies within an earthwork which used the steep slope of the Inferior Oolite scarp as a natural line of defence on one side and marks an even earlier phase of human utilization of this region which is discussed below (see p. 83). From the point of view of present land use it will be noticed that the Inferior Oolite, except on the steep portions of the scarp, is entirely used for arable cultivation, as is the Great Oolite to the East (see below, p. 101), but it is interesting to note the much higher proportion of grassland, not only on the Lower Lias Clays but also on the Marlstone ledge.

The study of rock strike on air photographs and the consequent elucidation of structures over large areas has been greatly developed by the geologists of the oil companies in order to trace the lay-out of oil-bearing formations, especially in the Middle East. Such large-scale studies are hardly possible, however, in Britain, but in individual areas the strike of the rocks can be a determining factor in the evolution of landscape and drainage and in land use, and Plate 12 is included to illustrate such an area in the Southern Uplands of Scotland. The rocks of the region are entirely of Silurian age, but their actual character has been modified in this locality, which lies in the aureole of contact metamorphism round the Cairnsmore Granite intrusion. In consequence, the hardening of certain elements in the rock succession has meant that heavy glaciation and later erosion have left groups of hardened strata upstanding as sharply defined hill areas and the effect of the original rock strike on the landscape has been greatly accentuated. The strike is clearly along the line indicated by the arrow $A-A'$ and the dip is almost exactly at right-angles to this line and in the direction of Black Loch in the top left-hand corner of the print. The relationship of relief in this area to the strike of the rocks is too obvious to call for further comment, but it should be noticed that the distribution of the isolated and rare patches of soil at B, B_1, etc., is entirely governed by the strike and consequent relief. It will be seen also that the predominantly transverse stream has developed short sections of its course along a

PLATES 1—16

Plate	Locality	Ordnance Survey Sheet no. 1:63,360	Latitude and Longitude	Sortie number	Print numbers
1	Bath	156	W 02°22' N 51°23'	106G UK 1522	6120
2A	Little Longstone Monsal Dale	111	W 01°43' N 53°14'30"	CPE UK 2598	4131
2B	Near Brent Knoll Somerset	165	W 02°55' N 51°16'	CPE UK 1869	4318
3	Kirkmirran South Scotland	Scotland 92	W 03°53' N 54°52'30"	106G Scot UK 42	4452
4	Comrie near Crieff	Scotland 63	W 03°59' N 56°23'	106G Scot UK 37	4038
5	Map of the area shown on Plate 4				
6	Braid Hills near Edinburgh	Scotland 74	W 03°12' N 55°55'	106G Scot UK 119 pt II	5154
7A	Photograph of Parallax Bar				
7B	Photograph of Wild Stereoautograph A5				
8	Photograph of Williamson-Ross SP3				
9A	Belstone near Okehampton	175	W 03°57' N 50°43'30"	CPE UK 2491	3362
9B	Menstrie near Stirling	Scotland 67	W 03°50' N 56°09'	106G Scot UK 120	4116
10	Near Wigtown South Scotland	Scotland 91	W 04°27' N 54°53'30"	106G Scot UK 42	4370
11	Horton Gloucestershire	156	W 02°20' N 51°33'30"	106G UK 1416	4380
12	Black Loch South Scotland	Scotland 87	W 04°21' N 55°01'30"	106G Scot UK 52	4276
13	Bratton Wiltshire	166	W 02°10' N 51°15'	CPE UK 769	3019
14A	Piddletrenthide near Dorchester	178	W 02°25' N 50°47'30"	CPE UK 1974	2371
14B	River Test near Basingstoke	168	W 01°15' N 51°15'	CPE UK 2102	4305
15	Cherhill near Calne	157	W 01°56' N 51°25'30"	CPE UK 1821	1070
16	Painsthorpe Wold near York	98	W 00°44' N 54°00'30"	106G UK 1313	4159

Plate 1

Plate 2

Plate 3

Plate 4

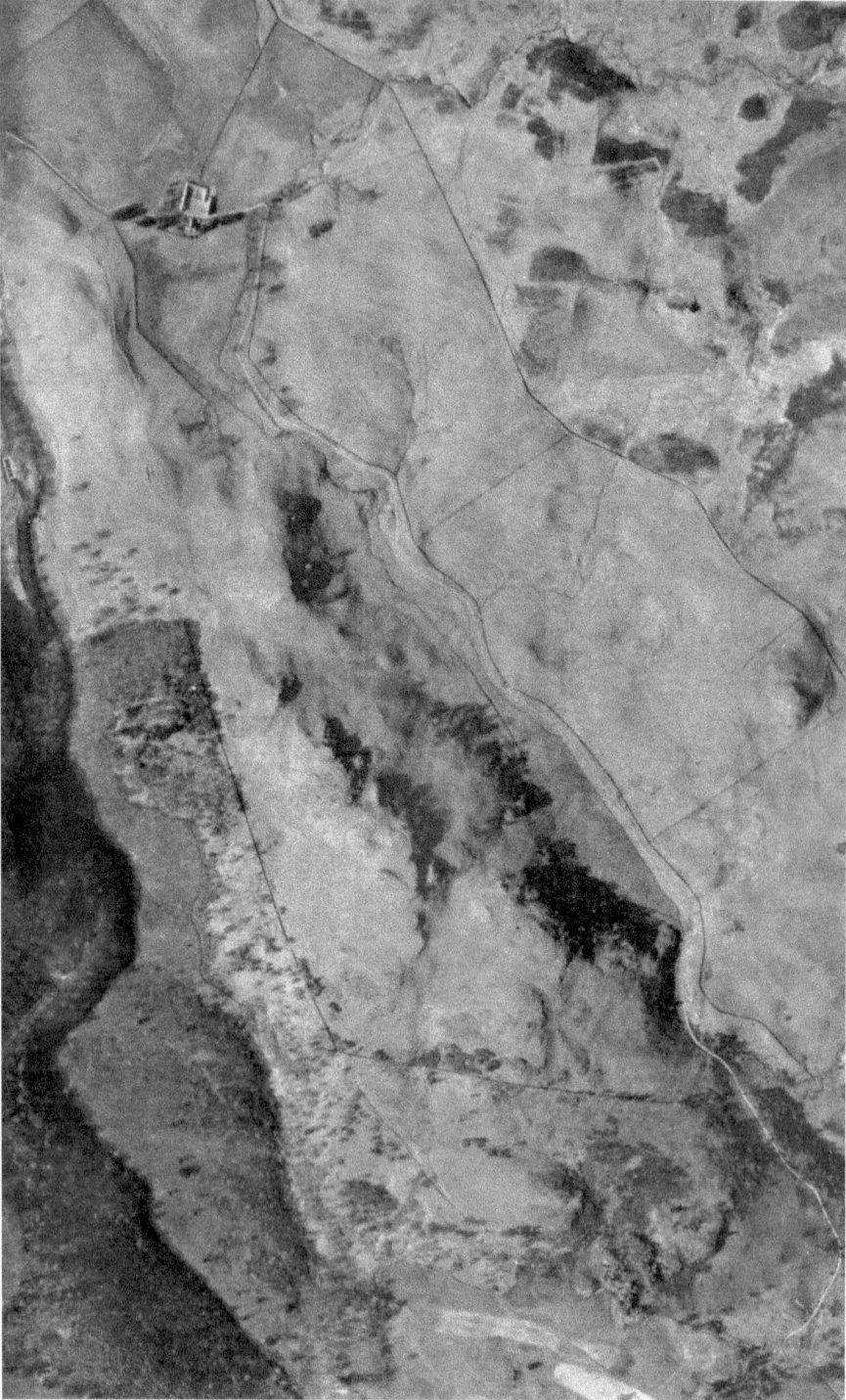

Plate 5

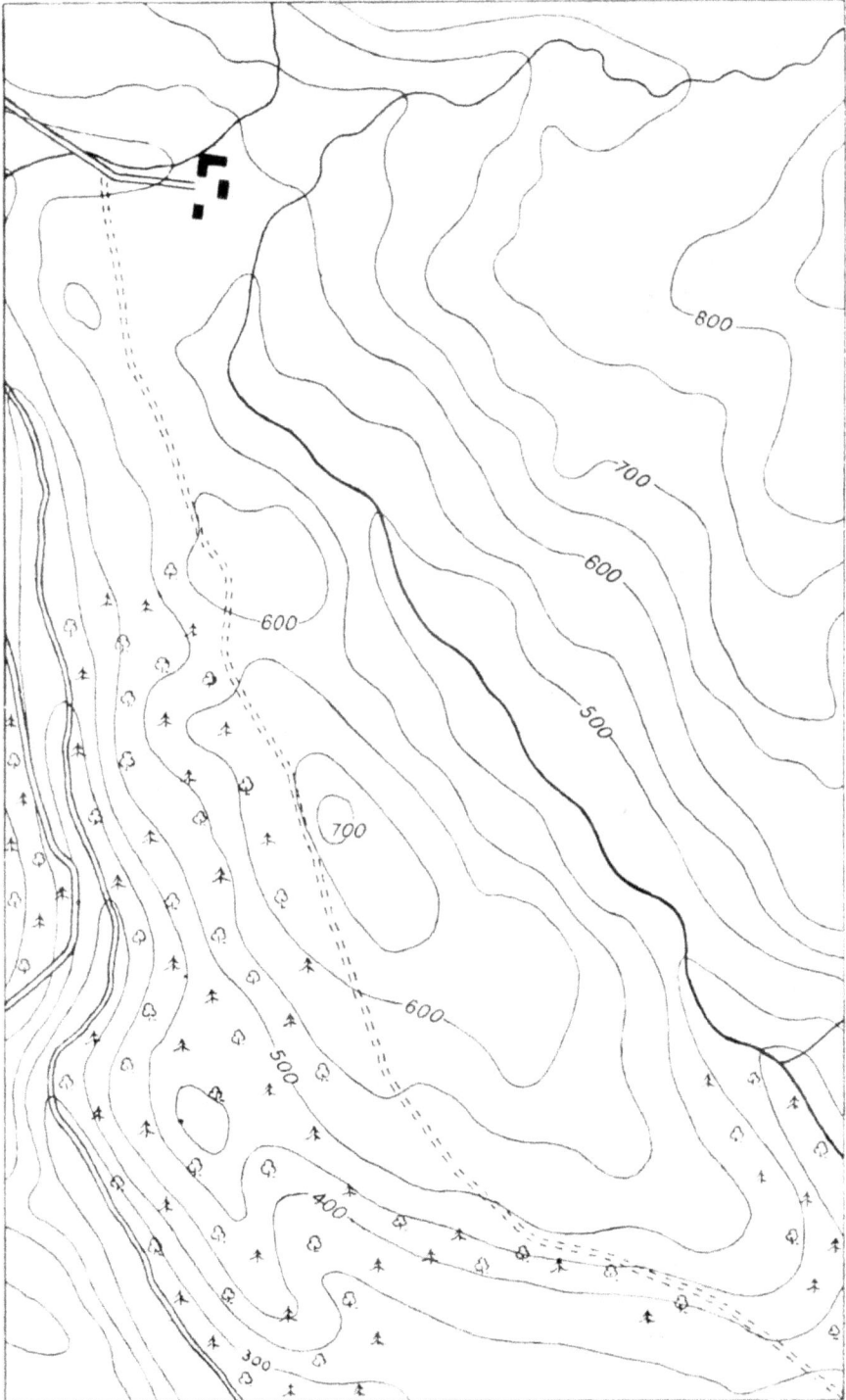

Plate 6

Plate 7

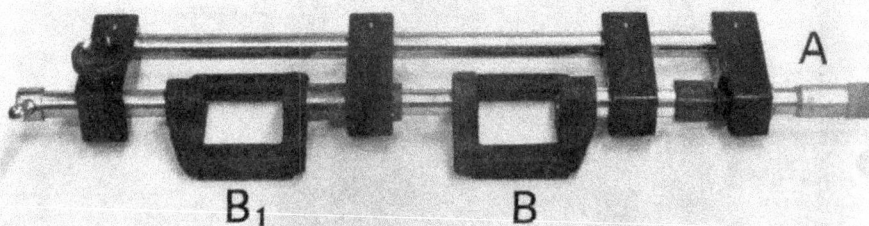

A

A

B₁   B

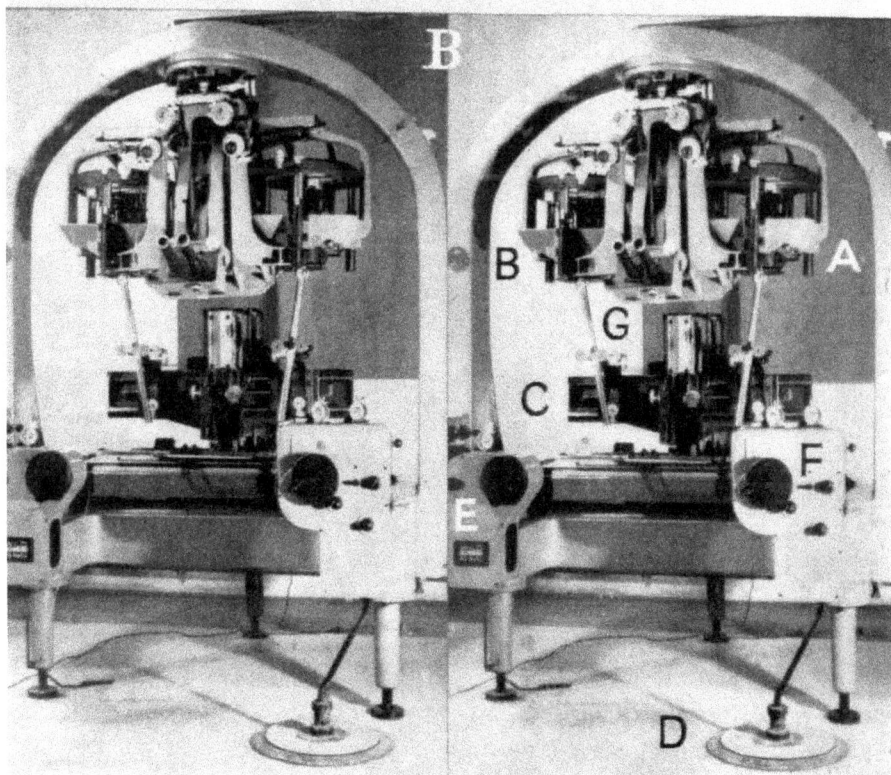

B

Plate 8

Plate 9

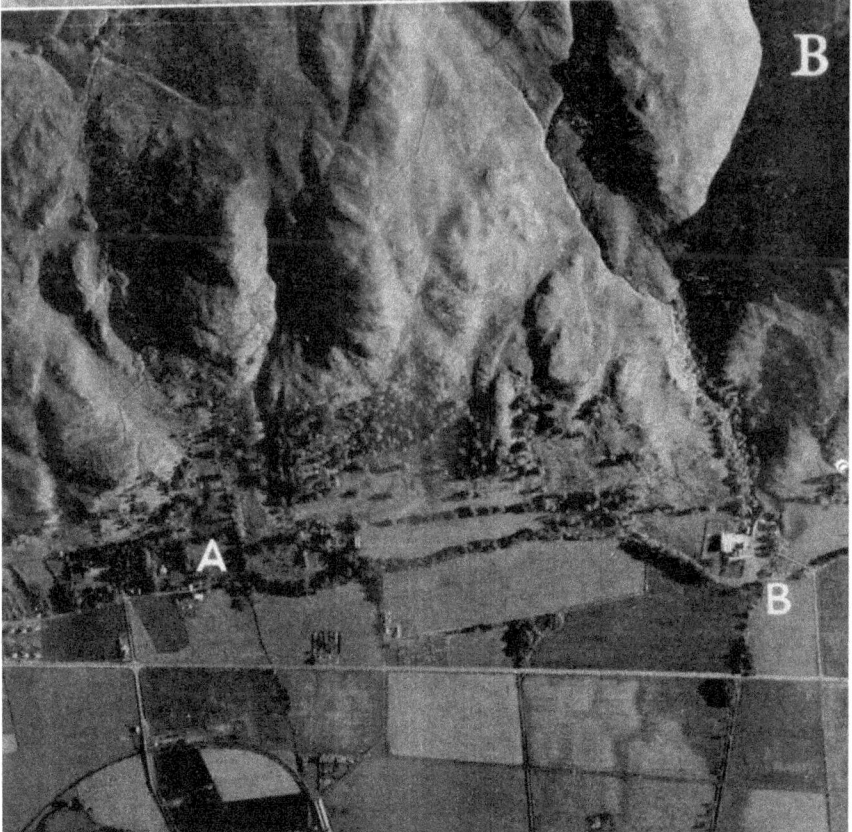

Plate 10

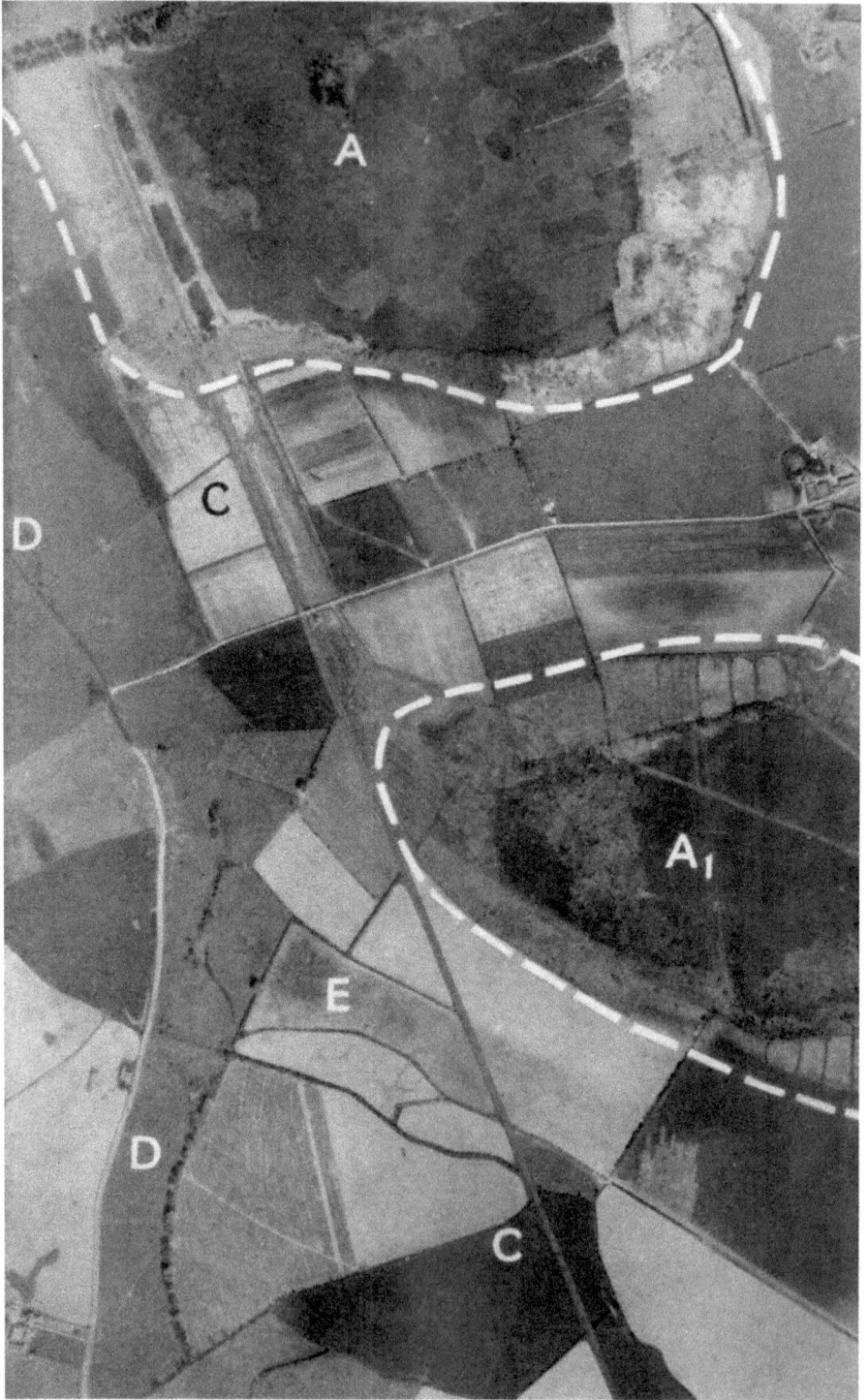

Plate 11

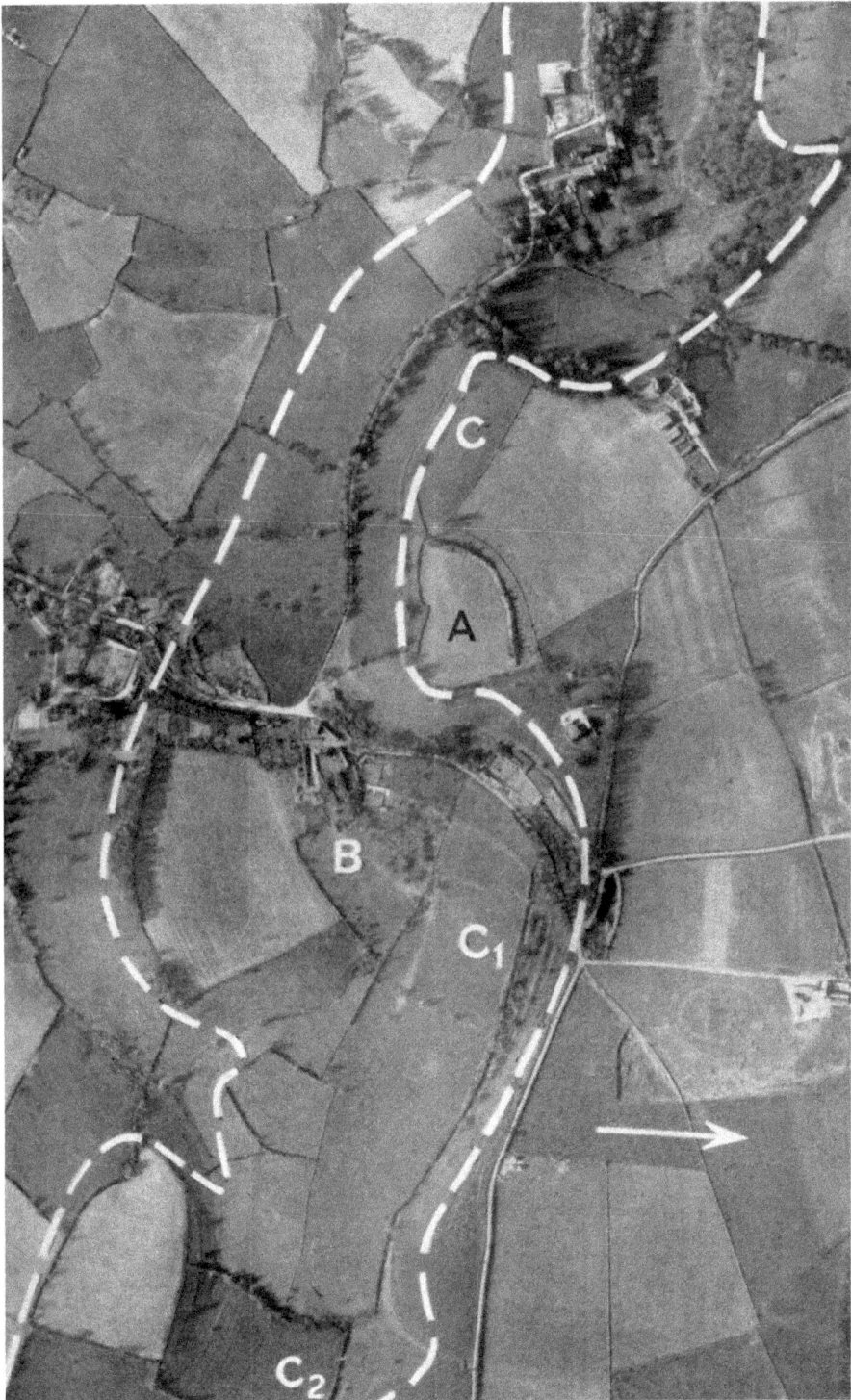

Plate 12

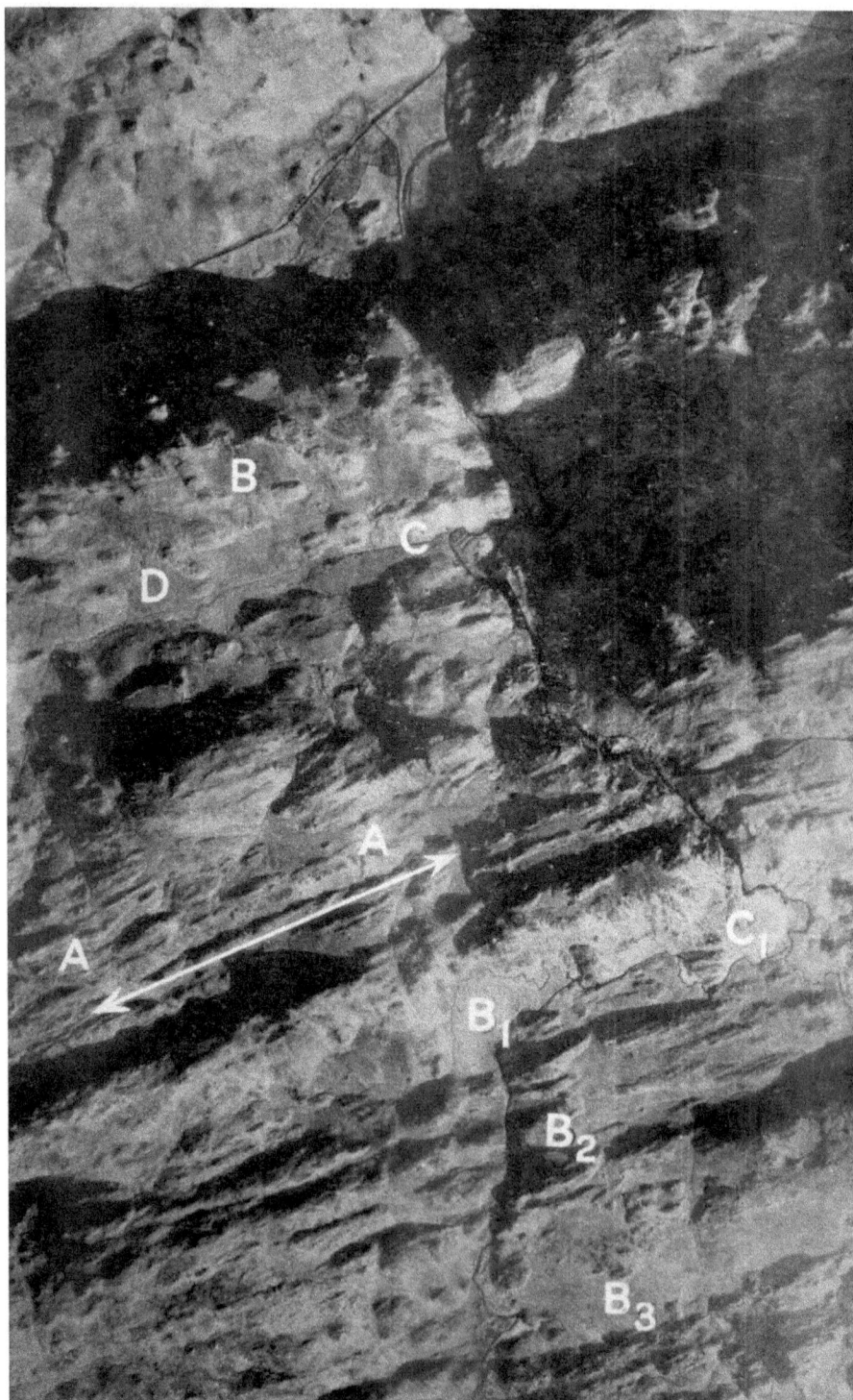

Plate 13

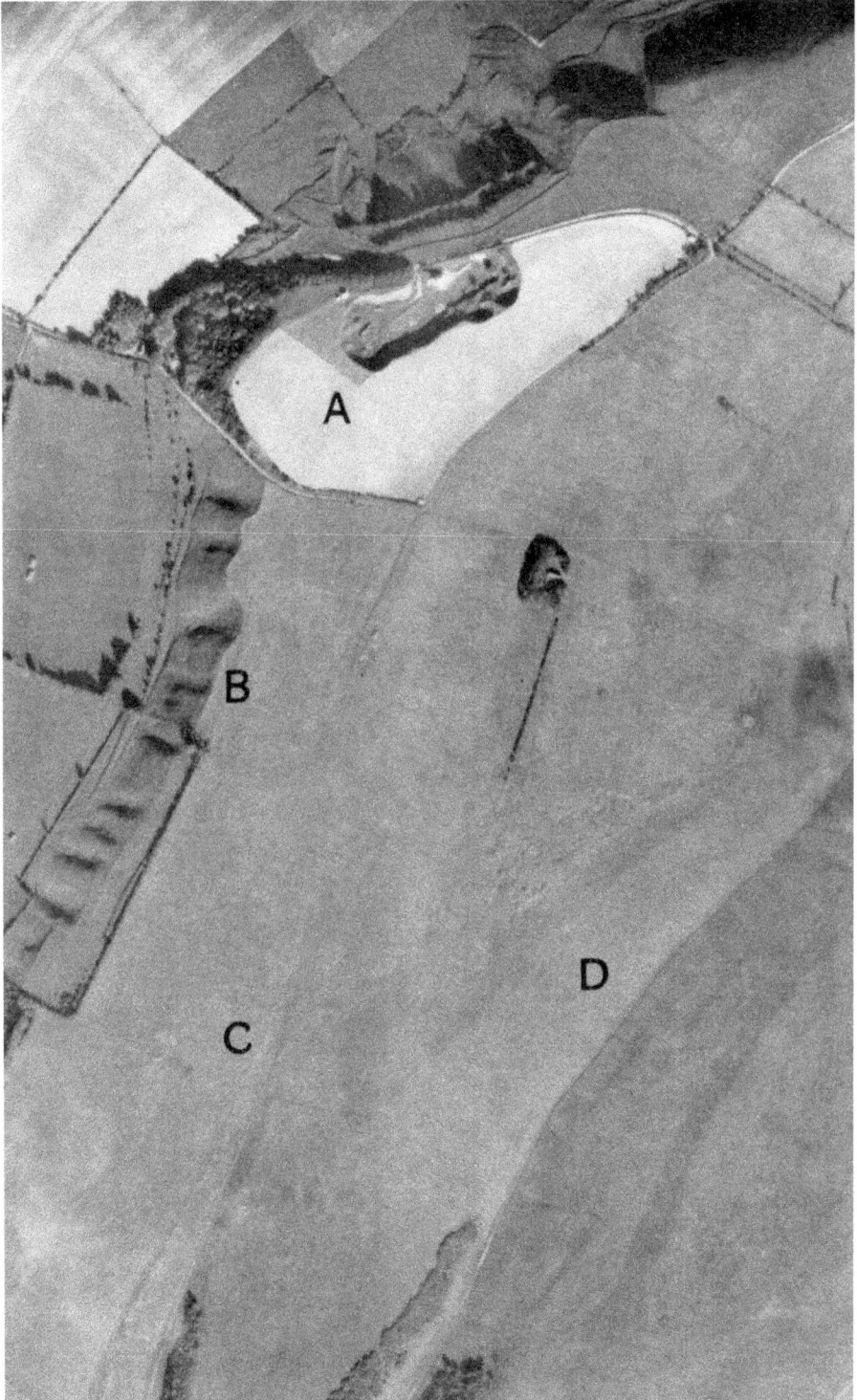

Plate 14

Plate 15

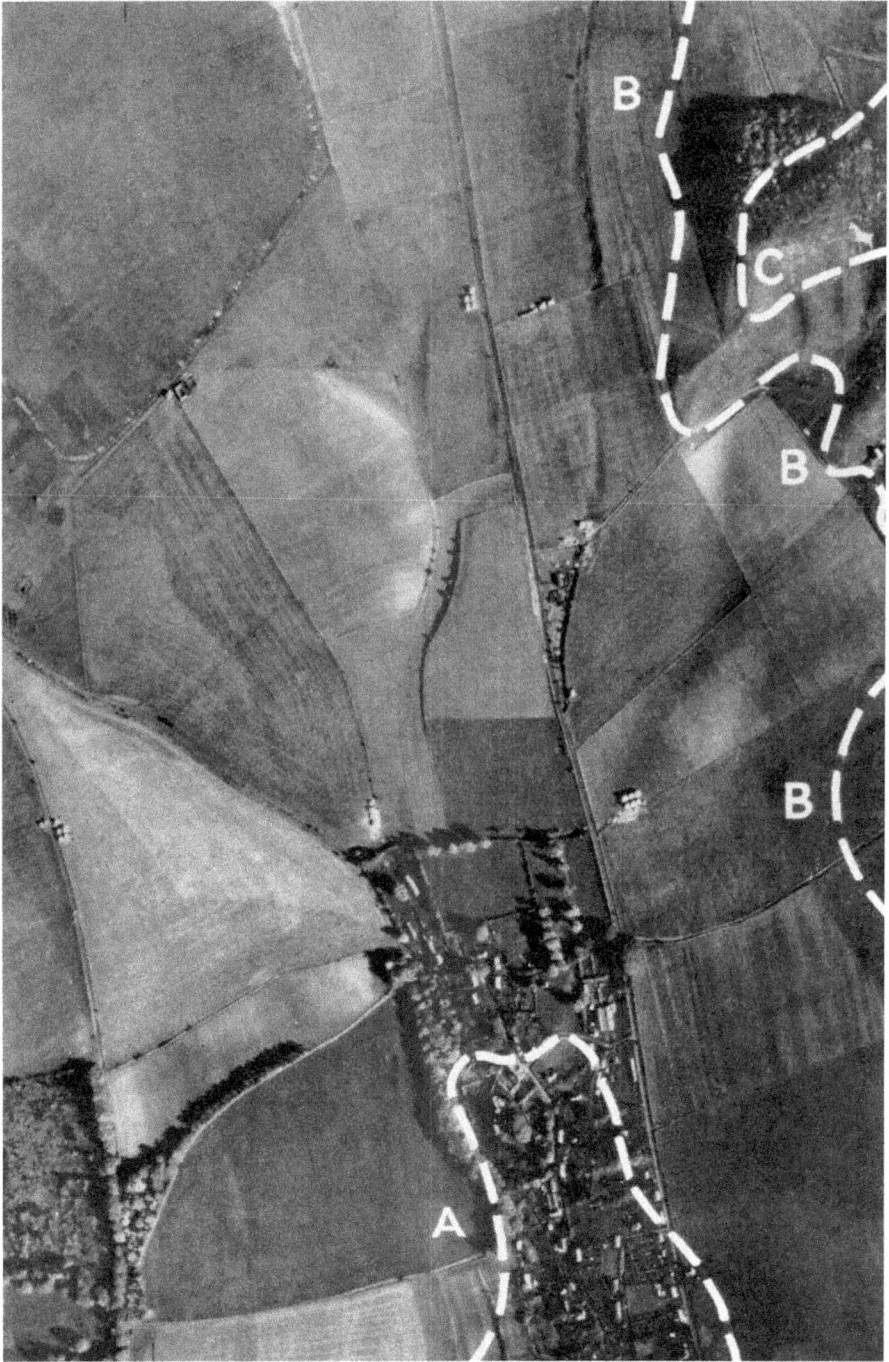

Plate 16

direction parallel to the strike at C, C_1 and further longitudinal drainage, modified by artificial cuts, is being developed, whilst Black Loch itself lies in a longitudinal depression parallel to the strike. Human utilization of the area is inevitably limited, but on three of the isolated areas of soil in the longitudinal depressions the existence of drainage channels shows that attempts have been made to use these soil pockets, and at D the long morning shadows show up the furrows in one such patch.

From a geographical point of view the clearest demonstration, by means of air photographs, of the effects of geological conditions can be seen where the existence of a particular type of rock leads to the development of a highly characteristic morphology and landscape, especially where it can be demonstrated too that there is a fairly distinctive and characteristic human response to the physical conditions. Accordingly, Plates 13 to 28 have been selected to cover a range of rock types which normally tend to produce such individualistic landscapes, though in most cases it has been necessary to introduce a number of examples to demonstrate very different surface phenomena in different districts lying on rocks which are at any rate nominally identical.

Chalk landscape is portrayed in Plate 13 in the form associated with the scarplands of Southern England, and the photograph shows a small portion of the Chalk scarp itself between Westbury (Wilts) and Bratton. Above the scarp the greater part of this portion of the Downs remains in the form of the gently rolling open grassland once characteristic of large areas of the Chalk, though even here it will be seen that recent changes in agricultural practice have resulted in one large field at A being put under the plough. Below the scarp the contrasting landscape of enclosed fields in varied agricultural use points to the narrow zone of the scarp itself as a dividing line from the point of view of human activity. Owing to the erosion of the scarp face the junction of the Chalk with the Upper Greensand lies a considerable distance in front of the scarp itself so that all the lower land shown on Plate 13 is on the Chalk, the top left-hand corner of the print being almost exactly on the junction with the Greensand. Nevertheless, the early development of agricultural settlement along the line of the Upper Greensand outcrop at Westbury, Bratton, Lavington, etc., meant that the whole of the area immediately in front of the Chalk scarp and between it and the Clay vale to the North is an area of old settlements and long-enclosed fields which straddle the junction of the Chalk and Upper Greensand, though in this case the chalky nature of the fields is amply evident on the photograph with its characteristic white and grey mottled effect on the arable fields.

The fact that the foot of the scarp and the land immediately in front of it has long been a zone of great importance for human settlement is emphasized on this photograph by the evidence of early cultivation patterns in the form of lynchets on the lowest slopes of the scarp face itself. Immediately below the markedly fluted portion of the scarp at *B* these take the form of long, narrow, terrace-like fields which are unfortunately picked out but partially obscured by the hedges. Farther along the scarp, however, just outside the zone of dense shadows cast by the hills, the pattern of old fields is sharply defined by the shadows of their own terraced edges.

The scarp itself is distinguished, particularly in the lower portion of the print, by its remarkably sharp crest and by the fluted appearance of the scarp face, erosional features which distinguish the West of England portion of the Chalk scarps, whilst the noticeably treeless character of the scarp over most of the area photographed is a common, but by no means universal, characteristic of this topographical feature in the West. The long, tapering, dry valleys usual in upland Chalk country, especially in the Wolds, are not common near the Western edge of Salisbury Plain, but two fairly good examples are to be seen in the lower half of Plate 13 at *C* and *D*, the larger one being made rather more obvious by the line of a footpath following the lowest part of the valley itself.

Plate 14 is intended primarily to show a highly contrasting landscape on the main area of the Chalk of Southern England away from the scarp edge and in particular to emphasize that the open Chalk downlands of the Western parts of Salisbury Plain are not by any means typical of land use throughout the central downlands of England. Photograph (*a*) shows a portion of the valley of the River Puddle at the village of Piddle-trenthide and the surrounding Chalk country. It will be seen that here the Chalk has been eroded into a series of rounded combe-like valleys branching from the main valley of the Puddle, land forms often regarded as being normal in the river erosion of Chalk areas. The sides of the tributary valleys and the main valley itself above the village are composed of the Middle (or so-called Hard) Chalk, whilst the crests of the intervening rounded spurs or Downs have a capping of the Upper Chalk, a distinction which appears to be evident on the photograph, though this is probably due entirely to the fact that the geological distribution coincides precisely with the relief forms. The village of Piddletrenthide lies on the Valley Gravels which cover the floor of the main valley from that point downstream, and the sharp contrast in settlement pattern, vegetation and land use between this broad, main valley and the surrounding countryside only serves to emphasize the true character of the general landscape. The side valleys in the Chalk are almost entirely devoid of evidence of surface water, though in one case there is an irregular white line which may represent a temporary stream bed. Nevertheless, though the area is thus entirely characteristic of a Chalk landscape from the point of view of relief and drainage, it should be noticed that from the point of view of land use it does not repeat the pattern of Plate 13 but is almost entirely covered with enclosed fields in arable cultivation. The only exceptions to this general rule are interesting in that the small areas of open grassland are invariably confined to the valley sides in the Middle Chalk where the slopes are steep and where apparently there is comparatively little soil cover.

The River Puddle itself is a perennial stream and has therefore been a reasonably satisfactory source of water supply from early times, and this fact, together with the light, well-drained but reasonably fertile character of the Valley Gravels, probably explains the existence of a band of settlement along the valley which can be traced on the map of the area (O.S. 1:63,360, Sheet 178) almost throughout its length. The photograph, particularly near the bottom edge, contains evidence of the very intensive cultivation of the valley floor in the patchwork of small fields, whilst the tendency, along perennial stream valleys in the Chalk country, for settlements to be strung out in long lines to make maximum use of the favourable conditions of the valley bottom is amply borne out by a study of the distribution of buildings along the portion of the valley shown on this print.

Photograph (b) shows the source of the River Test and is partly of interest for that reason, since it shows the characteristic upper portion of a Chalk stream with an insignificant valley which is abruptly enlarged with the addition of water from a number of springs near A, the point normally regarded as the source of the river. The main purpose of the print, however, is to demonstrate the remarkably rich arable landscape of this portion of the Chalk country and from this point of view a geological division shown on the photograph is of some interest. On the lower portion of the print, that is to the South of the River Test, the Chalk is overlain by the Reading Beds of Clays and Sands and though the distinction between this area and the Chalk to the North of the Test can be detected on the print in the slightly different soil colours in the ploughed fields, yet from the point of view of land use there is remarkably little contrast, and the photograph should serve to emphasize once again the fact that the Chalk, even without a covering of other material, does not necessarily produce open grass downland.

As a final example from the Chalk downlands of Southern England, Plate 15 shows an area of the Marlborough downs to the West of that town and including the White Horse. The village of Cherhill lies on the Upper Greensand which in this district has been exposed in a valley cut into the edge of the Chalk, but Cherhill is only one example of a series of such villages sited on the Greensand along the Western edge of the Chalk in Wiltshire—the outcrop of the Upper Greensand is outlined at *A*. The remainder of the print shows the characteristic rolling surface of the Lower Chalk with dry valleys converging on the valley at Cherhill, almost entirely devoid of trees and divided into very large fields by fences. At *B* and *C* on the edge of the print the outcrops of the Middle and Upper Chalk are outlined respectively, and it will be seen that they form areas of appreciably greater elevation which are also distinguished from the surrounding lower land by the fact that their steeper slopes are not used for arable cultivation. It is of interest to notice, however, that associated with the break of slope from the Lower to the Middle Chalk in the area between the White Horse and the main road there are a number of traces of lynchets following the general outline of the Middle and Upper Chalk hill.

Plate 16, the last of the photographs showing Chalk country, is intended to demonstrate some of the distinctive features of the Wold landscape and covers a small portion of Painsthorpe Wold in Yorkshire. From the point of view of physical geography the main interest in the photograph lies in the fact that it contains a nearly perfect example of the long, deep, tapering valleys so characteristic of Wolds scenery, though it should be remembered that the more rounded type of Chalk valley does also occur in the same locality. Unfortunately, owing to a highly peculiar optical effect produced by the direction of the shadows, the valleys on this particular print appear to many people to stand up in reversed relief as sharp-crested ridges, and this may prevent a proper appreciation of the true relief of the area which in such a case may only be understood by the use of a stereoscopic pair of photographs. The valleys are normally completely dry and taper out very gradually indeed into the surrounding Wold—the centre branch being particularly interesting in that its course can be traced across the nearly rectangular plantation and half-way across the adjoining field, where it is scarcely discernible even on the ground.

Much of the land on the Wolds has been brought under cultivation in comparatively recent times, a fact which is probably responsible for the rectangular lay-out of the field patterns, the long straight sections of the roads and the relatively small number of large farms. Despite the chalky character of the soil, which again produces the distinctive white and grey mottled effect on the photograph, the land is entirely under arable cultivation, once more with the only exception of the steep sides of the valleys. One of the main reasons which has retarded agricultural development on the Wolds has been the exposure of this upland area to very strong winds, and it is interesting to notice that the two large farms on the photograph have planted wind-breaks of trees on their Northern and North-eastern sides and in one case a field boundary has also been planted out to provide shelter from the Easterly winds.

Plate 17, the first example of the landscape associated with the Carboniferous Limestone, covers part of the Southern limb of the Limestone at the Western end of Mendip near Shipham where the Limestone forms a Northward-facing scarp and dips Southwards as may be seen on the West side of the through valley. The Lower Limestone Shales, in this district, only form a very narrow belt at the foot of the scarp itself and are not identifiable as a separate relief feature, so that the greater part of the print is made up of the Limestone proper. The narrow zone of lower land at the top of the print and the area of lowland below the woods to the South lie on the Trias which surrounds Mendip.

The very thin soil cover on the Limestone can be appreciated from the appearance of the Limestone at A and the slight evidence of solution phenomena in the fields near B. It will be noticed that the whole of the relatively level area of the Limestone is grass covered with isolated patches of scrub and with some poor woodland on the Southern slopes, and there is evidence that even in the enclosed fields much land has reverted to what is little better than rough grazing. Most of the field boundaries are marked by the straight stone walls which are characteristic of the Limestone uplands, though recently erected field boundaries between A and B are of modern post and wire construction. The three large quarries are, of course, highly characteristic of Limestone areas, and it is of interest to notice that their location enables them to make use of the line of communication provided by the Cheddar–Shipham road passing through the valley.

Even the small areas of the Trias appearing on this print serve to demonstrate the distinction between the Limestone uplands and the Triassic lowlands surrounding Mendip with their tree-lined field boundaries and, particularly to the South, their much greater extent of arable cultivation. In the extreme bottom right-hand corner of the print it will be noticed that the arable cultivation has produced a narrow banded effect on the photograph of the fields, an effect which is often an indication of a market-gardening type of land use with numerous narrow belts devoted to different crops within a single field. In this area a certain amount of market-gardening is carried on, but most of the strips are in fact strawberry 'fields', a very common form of land use along the Southern edge of Mendip, where the suitability of the Triassic soils and their well-drained character on the fairly steep slopes below the Limestone combine with a Southern aspect and considerable shelter provided by Mendip to produce perfect conditions for strawberry culture.

Plate 18 consists of a mosaic of three photographs (reduced in scale) to show the Carboniferous Limestone of the Northern limb of Mendip at its Western end near Burrington, though in this case other important rocks enter into the area on the photographs. The Carboniferous Limestone itself forms the broad belt of rounded hills which are the true edge of Mendip on the North and are marked *AAA* on the print, the continuity of the hill mass being broken at *B* by the deep cleft of Burrington Combe. Immediately to the South of the Limestone hills there is a relatively smooth and lower zone marked *CCC* on the print which marks the outcrop of the Lower Limestone Shales, an element in the rock succession which is much more important here from the point of view of relief than in the area shown on Plate 17. Farther South the area marked *DDD* and bounded by the dotted line to the West, is distinguished on the photograph by the banded effect produced by shadows cast from slight ridges occasioned by the bedding of the rocks. This is the area of outcrop of the Old Red Sandstone which forms the 'core' of Western Mendip, and its unusual appearance on the print is highly characteristic of most air photographs of this particular Sandstone. The remainder of the area shown on Plate 18, that is to the West of the dotted line and to the South of the Limestone and also the narrow belt of lowland to the North of the Limestone hills, lies on the Keuper Dolomitic Conglomerate.

The Limestone hills again have a characteristic vegetation of rough grass and scrub, though their steep Northern slopes have been successfully planted with impressive woodlands, and once more it is possible to appreciate the thinness of the soil cover. For example in many places near the earthwork at the Western end of the hills the faint straight lines mark the edges of the strata and the large number of surface excavations indicate that only the shallowest of trenches was needed to expose the rock in the continual search for lead, iron and ochre which went on in this area during the Middle Ages and indeed up till the beginning of the twentieth century.[1]

There is evidence of the porosity of the Limestone at *E* where the pitted appearance on the photograph indicates the presence of a considerable number of swallets, but perhaps the most striking evidence is that afforded by the stream *F*, flowing from the Old Red Sandstone and across the Lower Limestone Shales which disappears underground near the edge of the Limestone at the entrance to Read's Cavern (*G*). Farther to the

[1] Gough, *Mines of Mendip* (Oxford, 1930), p. 244.

East the two parallel streams flowing across the Old Red Sandstone from Blackdown enter the Limestone where their valleys join the middle section of Burrington Combe. No final proof has yet been given that Burrington Combe represents the course of a collapsed water-worn cavern, but if such a theory were accepted the first-mentioned stream (F) and the passages of Read's Cavern might well be regarded as representing a similar phenomenon in an early stage of development—an assumption made all the more tempting by the ease of appreciating on the photograph the many remarkably close similarities which do exist.

The open grassland character of the Limestone uplands made them important areas of settlement and communication in prehistoric times, and Plate 18 includes two examples at H and J of hill forts or camps which demonstrate the importance of this Limestone portion of Mendip in early times. It is significant that both are situated at points which command valley routes through the Limestone hills. The camp at H commands the lower part of Burrington Combe, but judging from its small size and unimpressive earthworks it was not considered a site of great importance, possibly because the narrow cleft of Burrington would itself be easily defensible and also because the route up the Combe merely leads on to the central part of Mendip and could scarcely be regarded as a major line of communication. On the other hand, Dolebury Camp (J) is a most impressive earthwork, and this fact may well be related to the circumstance that the valley immediately to the West of the camp is much more easily traversed than Burrington Combe and in addition this route, followed now by the main Bristol–Bridgwater road, formed an important and easy link between the lowlands to the North and South of Mendip.

The contrasting land utilization of the lower areas on the Dolomitic Conglomerate is largely apparent on the photograph because of the pattern of enclosed fields, but it should be noticed that here the majority of the fields are in grass, since this is part of the West of England dairying region, and the special local conditions which accounted for the arable appearing on Plate 17 do not apply. The remarkable pitted appearance on the photograph at K is caused by the remains of calamine (zinc ore) workings. Comparatively little lead was worked in the Dolomitic Conglomerate round Mendip, but large amounts of calamine were extracted, most of it occurring within four or five feet of the surface so that it was possible for small groups of miners to work the ore over large tracts of land, and at various times the entire working population of the near-by villages of Rowberrow and Shipham were engaged in this occupation.[1]

[1] See Gough, op. cit., passim under 'Rowberrow'.

The most striking topographical feature shown on Plate 19 is undoubtedly the valley form of Monsal Dale which shows the abrupt
changes of direction normally associated with such valleys in the Carboniferous Limestone of Derbyshire, probably because of the effect of the
direction of jointing in the Limestone upon the pattern of river development. In particular the sharp angle of the crest of the valley slope at *A* is
highly characteristic. The cross-section of the valley is also typical of the
Limestone dales with its great depth (approximately 400 feet below point
A) and its extremely steep slopes, and it should be noticed that these slopes
in all cases are either wooded or covered with a rough scrub-like vegetation, the latter being more common where the light appearance of the
photograph reveals that there is comparatively little soil cover over the
underlying Limestone. The depth of the valleys naturally presents serious
problems in the development of communications, and, whilst the main
road uses the valley floor, the railway near *B* has been designed in such
a way that its course lies in a cutting half-way down the valley slope, then
crosses the valley by means of a viaduct and enters a tunnel in the opposite
valley face, a form of construction repeated at intervals throughout the
course of this line where it traverses the Carboniferous Limestone.

Outside the valleys the Limestone in this area presents a relatively
uniform surface, and in further contrast with the parts of Mendip shown
on Plates 17 and 18 practically the whole of this area is covered with a
patchwork of quite small enclosed fields, the fine, straight white lines
caused by the Limestone walling of the field boundaries being a distinctive
feature of air photographs of this region. Practically all the fields shown
on the photograph are, however, laid down in permanent grass or very
long grass leys, the lighter appearance of some of the fields being probably
due to the fact that they have been recently cut or rollered.

The economic importance of the Limestone itself may be deduced from
the presence of a quarry at *C*, the site of which enabled it to use the railway
on the valley slope immediately below as its normal route for shipping
out the material. The line of pitted ground immediately above the quarry
represents the site of former lead workings which again are common
throughout the Limestone in Derbyshire.

Settlement is noticeably very sparse throughout the area on the photograph and takes the form of isolated large farms which themselves often
have their outbuildings dispersed at points away from the main farm
buildings to minimize the movement of stock and materials. One of the
main problems associated with the operation of these farms arises from the
absence of surface streams on the upland Limestone and the great depth of
the valleys, a problem which was overcome by the use of artificially

constructed circular ponds which are to be seen at many points on the photograph at DDD, etc.

As in the case of Mendip the upland areas of the Limestone proved suited to the requirements of man in prehistoric times and at E the outlines of a camp can be identified—one whose position at the crest of the extremely steep valley slope almost justifies its description as a promontory fort.

In the area FF_1F_2 it will be noticed that a zone of land is distinguished by a slightly rougher surface and by the fact that it has not been enclosed, circumstances which appear to arise from the presence of an intrusion of Dolerite along this side of the valley.

Plate 20 shows the upper portions of the adjacent and aptly named Coombs Dale, but is included primarily to show the long line of Rakes A–A_1 and B, the former known locally as High Rake at A and Deep Rake at A_1. The name 'rake' was given by the lead miners of Derbyshire to the elongated series of cavities with mineralized content which follow the major joint planes of the Limestone and which were traced by the search for lead in the surface workings and later shaft mines, so that the natural ridging of the surface along the line of cavities has been accentuated into an appreciable trough running cross-country in the manner shown on the photograph. The white appearance of the rakes is occasioned by the removal of vegetation and soil in the lead workings which leaves the Limestone exposed at the surface. At various other points on the photograph, such as C and D, there are local traces of lead workings but the great majority of them are concentrated in a broad belt EE corresponding to the slope of Longstone Edge where the edges of the Limestone facilitated surface and adit working. The line of Longstone Edge is of interest too because its base marks the junction of the Carboniferous Limestone with the Lower Limestone Shales which form the area in the bottom portion of the print at F, a distinction which is responsible for the rather different land utilization and landscape in the latter lower area with its tree-lined road and slightly greater evidence of arable cultivation.

The photographic quality of Plates 21 and 22 is less satisfactory than that which has been obtained for other plates, a fact which is almost entirely due to the difficulty of obtaining brightly illuminated, clear photographs over the Northern part of the Pennines, where the liability to low cloud often produces conditions comparable to those evident in Plate 21. This photograph shows the portion of Gordale Beck immediately above Gordale Scar and the waterfall. It serves, of course, to illustrate again the characteristic erosional feature of sheer-sided gorges in the Carboniferous Limestone, the deep shadow unfortunately hiding in part the marked asymmetry of this particular example. Despite the poor quality of the print, it does, however, give a reasonably clear impression of the surface appearance of the Great Scar Limestone and of the almost complete absence of soil in the region surrounding the beck so that the contrast between the upland Limestone of the North Pennines and that of Derbyshire or Mendip may be appreciated. Another feature of the surface erosion of the Carboniferous Limestone which is common in the Malham–Gordale area of the North Pennines, but somewhat rare farther South, is the development of grykes or clints. This phenomenon results from the erosion and solution of the Limestone along a very large number of adjacent joints, producing a pitted surface resembling a pavement of large slabs with wide and deep joints separating them, an appearance possibly responsible for one other local name for the surface—a 'street'. At *A* on Plate 21 a small area of grykes can just be discerned, but again poor photographic quality precludes a proper appreciation of its character.

The greyness of a dull day tends to accentuate the bleak character of the Western edge of the Cross Fell portion of the Pennines which is shown on Plate 22. The scars at the top of the photograph at *A* are part of Long Fell near its junction with Roman Fell, both of which form the precipitous crest of the Cross Fell edge overlooking the upper part of the Eden Valley. The position of the main mass of the Scar Limestone can be seen on the print, but it is not possible to determine the precise location of the Whin Sill which lies on top of the Melmerby Scar Limestone in this area and is responsible for the precise character of the scar line. The vertical streaks at *BBB* are the sites of former calamine workings which have to be approached from the lowland of the Eden valley by the track which zigzags up the steeper part of the Edge.

Except in points of detail, conditions on the Millstone Grit of the central parts of the Pennines are much more uniform than on the Carboniferous Limestone so that only one photograph of the Grit area has been included. Plate 23 shows a very typical example of the rounded bleak moorlands of the Millstone Grit near Todmorden, almost entirely devoid of agriculture, settlements or roads except in the main valleys and in the small 'intake' fields on the lowest edges of the moor. On the moor itself there are two major changes of slope, one along the dotted line *AA* at a height of about 900 feet, below which the edge of the moor drops sharply to a through valley (discussed below, Plate 96) which has been eroded across the main mass of the Grit. The second dotted line *BB* is at a height of approximately 1,000 feet, above which the ground rises much more steeply to 1,250 feet on Inchfield Moor (*E*), a break of slope marked on the photograph by the large number of small gullies above this line.

This minor relief feature marks exactly the junction between two different sections of the Millstone Grit, the Haslingden Flags on Inchfield Moor and the Middle Grit below the line *BB*.

Perhaps the best-known characteristic of the Millstone Grit is its value as a source of soft water, and at *C* and *D*, Plate 23 shows two reservoirs supplying water to Todmorden and Rochdale respectively, the circular object below reservoir *C* being part of the filtration plant. The precise location of the two reservoirs is interesting since it will be seen that the dams have been built across the valleys of two small streams where they have been cut back into the steeper slopes of the Haslingden Flags. The valleys below the reservoirs contain the only tree growth in the whole area, though even here it is noticeably less important than in the Carbon-iferous Limestone valleys.

One example of landscape associated with Sandstone has already been referred to in Plate 18 which shows a portion of the Old Red Sandstone area of Mendip. In the case of the Triassic Sandstones which form the Midland plateau and much of the lower land of Lancashire and Cheshire, however, extensive drift cover tends to prevent the appearance of specially distinctive landscapes, and Plate 24 has been selected since it covers an area where the outcrop of the Trias does produce important topographical consequences. The greater part of the print shows the plain of Western Cheshire with Bunter Sandstones and Bunter Pebble Beds underlying the drift, the level of the plain being generally between 200 and 250 feet. Ten miles South-east of Chester, however, the Bunter is replaced by the appreciably harder Keuper Sandstone which produces the extremely important range of hills from Bickerton Hill running North-eastwards to the isolated small hill on which Beeston Castle is situated. This last-named hill, made even more startling in appearance by the long morning shadows, appears on Plate 24 at *A*—it may be noticed that the position and even the shape of the castle itself is revealed in the shadow cast on the plain. Immediately to the South-west of Beeston there is the much longer range of the Peckforton Hills, whose presence is revealed on the photograph only by the fortunately long shadows. The average height of the Keuper Sandstone hills is between 500 and 550 feet, though at one point they rise to 695 feet. To the South-east of the range of hills the Sandstones are replaced by the Keuper Marls, but though there are minor differences of land use on the Marls their land-scape closely resembles that of the Bunter to the North-west and in con-sequence the range of Sandstone hills stands out from the general level of the Cheshire plain as a most remarkable relief feature, especially as they have weathered to form extremely precipitous slopes, particularly on the

West and North-west, unfortunately hidden in the shadows on Plate 24. One consequence of the isolation of this hill range has been that it has provided important strategic sites since prehistoric times, and on the crests of the hills there are examples of defensive sites from the Iron Age hill fort of Maiden Castle at their Southern end to the twin medieval castles of Peckforton and Beeston in the North, the site of Beeston commanding the valley of the Gowy which can be seen to the North with the Shropshire Union Canal and the main Crewe–Chester–North Wales railway line following its course through the Keuper hill area.

It is not proposed at this point to discuss the land use of this portion of the Cheshire plain since a photograph of a near-by area is to be described later as an example of a grass-farming region.

The first example of the Jurassic rocks of Britain, Plate 25, shows the scarp and crest of the Cotswolds immediately East of Chipping Sodbury and covers the succession from the Lower Lias Clays to the Fuller's Earth with the only exception that the Upper Lias Clays do not appear in this section. The junction of the Lower Oolite and the Fuller's Earth along the line AA' produces only an imperceptible break of slope as does the junction of the latter with the Great Oolite to the East (not shown on the print), so that in this part of the Cotswolds it is the outcrops of the Lower Oolite and the Midford Sands, defined by lines BB' and CC', which are associated with the scarp itself. Above the line of the scarp face (approximately BB') it will be appreciated that the surface of the Lower Oolite and the Fuller's Earth is virtually horizontal and forms a plateau-like area rendered even more uniform by the absence of trees, the low, dry stone walling and the universal arable agriculture. The scarp slope itself, though steep, is low and comparatively narrow, between lines BB' and CC', and only forms a really impressive relief feature round the small re-entrant valley near E. Nevertheless, it will be noticed that the boundaries of the large fields on the Inferior Oolite mark the crest of the scarp face, the slope of which at E, F and G is emphasized by a vertical arrangement of lynchets, the group near G being particularly well preserved.

The broader belt between lines CC' and DD' is the area of outcrop of the Middle Lias Marlstone which forms a nearly level platform at the foot of the main scarp, and throughout the whole of the Central and Southern Cotswolds this Marlstone ledge is the site of the majority of the villages and towns in the vicinity of the scarp. In this photograph the concentration of settlement on the Marlstone at the village of Old Sodbury is still apparent despite the fact that the settlement has spread on to the lower slopes of the scarp and on to the Clays below the ledge as a result of the tendency for building to follow the line of main road which crosses the

Marlstone and curves up the re-entrant valley to the top of the Cotswolds. The contrast between the dip slope of the Inferior Oolite and Fuller's Earth on the one hand and the scarp face and Marlstone ledge on the other is partly, therefore, a matter of settlement distribution, but, in addition, there is also the great distinction between the large, often rectangular, arable fields of the former area with their dry stone walls, and the much smaller, irregularly shaped fields of the latter with their mixture of arable and permanent grassland, their tree-lined hedges and the isolated patches of woodland.

The small area to the left of the line DD' lies on the Lower Lias Clays, but it is only in the top left-hand corner of the print that there is a sufficiently large area for one to see an example of the poor grassland that covered most of these heavy, ill-drained Clays until very recent times.

The small tower and earthwork at H is a ventilation shaft and spoil heap which marks the line of the Badminton Tunnel through the Cotswolds carrying the railway from South Wales and the Severn Tunnel to Swindon and London.

It would be misleading to regard the conditions shown on Plate 25 as typical of the whole of the Cotswolds. Reference to a small-scale map will show that Northwards from this area the Cotswolds increase in height, the land use changes, and the surface is broken to an increasing extent by wide and deep valleys cut into the scarp face and draining towards the Severn, the most impressive group of which centre on the great Stroud-water valley system. It is suggested, therefore, that Plate 67 be consulted at this point, though it is included among later photographs for other purposes, since it shows a portion of the valley of the Frome (known as the Golden Valley), about two miles upstream from Stroud.

Spot heights have been marked at intervals across the main valley in order to give a clue to its dimensions, and it will be appreciated that it is both deep and wide, contrasting noticeably with the valleys shown on the photographs of Monsal Dale in that it is appreciably more rounded in section and more nearly approaches the usual conception of a mature valley.

The Lower Oolite forms the whole of the higher land in this photograph above the line AA' on the West and above the line EE' on the East. In the East, above the villages of Thrupp, the Oolite produces conditions not dissimilar to those farther South, though with an appreciably less uniform surface and a good deal more woodland. In the West, however, the Oolite coincides almost exactly with the limits of Rodborough Common where there is open grass of a downland type which is often incorrectly regarded as being typical of most of the crest of the Oolite scarpland. Such patches of open grassland are, nevertheless, still fairly common in this portion of Cotswold, but they are confined, in the main, to areas in the immediate vicinity of the main scarp and to patches of upland, like Rodborough Common, which are surrounded on three sides by the deep valleys of the scarp streams. It is also characteristic of this heavily dissected portion of the Cotswold Upland that there are considerable patches of rich woodland on the lower slopes of the Oolite.

Below the outcrop of the Oolite the sides of the main valley, above the line BB' on the West and above the line DD' on the East, are composed of the Upper Lias Sand, a formation of great importance in this district from the point of view of settlement as will be seen later (see below, pp. 75–6). It will be noticed, too, that the small, rounded, side valleys are also developed in this formation and are all significantly devoid of important traces of surface water.

The Upper Lias Clay, absent from the district shown on Plate 25, reappears in this area and forms the floor of the main valley between the lines BB' and DD'. Here it is of extreme importance since the outcrop of this formation and the appearance of surface water has been inextricably linked with many of the more important aspects of the human geography of the Stroud area which are discussed below (see pp. 75–6). The small V-shaped area included within the line CC' marks the outcrop of the Middle Lias Marlstone and Northwards and Westwards from here the Frome has eroded the whole of its valley through the Cotswolds down to this formation so that its outcrop widens into a small plain on which the town of Stroud is situated.

Plate 26 consists of portions of four photographs made into a mosaic and reduced in scale to show a fairly lengthy portion of Lincoln Edge to the South of Lincoln itself. Here the Lower Oolite, which covers the whole of the print to the East of the line AA', has a landscape comparable with the dip slope of the Cotswolds shown on Plate 25. The land is almost entirely devoted to arable cultivation, the fields being divided by low stone walls or even merely by shallow ditches so that an exceptionally uniform appearance results. The line FFF of field boundaries, narrow lanes and short stretches of road marks the alignment of the Roman road of Ermine Street running along the high ground Southwards from Lincoln, and it has been suggested that the remarkably rectangular arrangement of field boundaries in this area arises from the fact that the creation of the field units on either side of the straight line of the Roman road naturally tended to take the form of straight-sided fields based on it. The Roman road has now been replaced as a main line of communication by the scarp road from Lincoln to Sleaford, farther to the East. The former Roman road is therefore only used locally along short stretches, and at G there is an interesting demonstration of the changed importance of roads in this region, since the transverse road from Harmston (C) to the Sleaford road, which formerly used a portion of the North–South Ermine Street, has in recent years been curved to avoid the right-angled bends which occurred on joining and leaving Ermine Street.

It will be noticed that the crest of the scarp which coincides with the outcrop of the Inferior Oolite along the line AA' also coincides with a long line of field boundaries, probably because of former differences of land use on the scarp face and the plain to the West. The three villages of Harmston (C), Coleby (D) and Boothby Graffoe (E) are situated exactly on the line of the geological junction and were ideally situated, therefore, as centres of settlement associated with the cultivation of the Oolite soils of the dip slope to the East and the Clays and Marls of the scarp face and vale to the West. The highest area of the Oolites is a narrow belt of land a little over 250 feet in height parallel with the Western edge of the outcrop so that the Lincoln–Grantham road (HH) is in fact a true ridge road, and it seems extremely probable that it lies along the line of a pre-Roman crest trackway which is thought to have followed the line of the open land on the Jurassic scarp across England from the Cotswolds to Lincoln and beyond, at any rate in the later Iron Age, and which certainly did exist in this Lincoln Edge portion of the scarp.[1] The identification of the modern road with the alignment of a prehistoric trackway along the crest of the Edge may explain the fact that the three villages shown on the photograph and others to the North and South lie off the line of the road itself. These later settlements of Saxon and in some cases Danish origin appear to have been deliberately sited off the actual crest of the hills at the point where the different soil types of the Lias and the Oolites met, but communications between them were maintained along the line of the ancient routeway which has subsequently become of much greater importance than the purely strategic Roman road largely because it did in fact function as a connecting route near to the long line of villages following the Western side of the Edge.

West of the line AA' and on the scarp face the Upper Lias Clay appears in a comparatively narrow belt to be succeeded by the Middle Lias to the West of the line BB'. The outcrop of the impermeable Upper Lias Clays results in the appearance of a series of springs at its junction with the Oolite along the line AA', and the water supply provided by these springs

---

[1] Evidence, mainly in the form of finds of Iron Age material along the line of the trackway, is more conclusive in this section than in the Cotswolds and Northamptonshire.

Plate	Locality	Ordnance Survey Sheet no. 1:63,360	Latitude and Longitude	Sortie number	Print numbers
17	Near Shipham Somerset	165	W 02°48′ N 51°18′	CPE UK 1869	4289
18	Burrington Somerset	165	W 02°46′ N 51°19′30″	CPE UK 1869	3278 3280 3281
19	Monsal Dale	111	W 01°45′ N 53°14′	CPE UK 2598	4133 4135
20	Longstone Edge near Bakewell	111	W 01°40′ N 53°15′30″	CPE UK 2598	3098
21	Gordale Beck	90	W 02°08′ N 54°04′30″	106G UK 1514	3023
22	Long Fell near Kirby Stephen	84	W 02°22′ N 54°34′	541/108	4128
23	Inchfield Moor near Todmorden	95	W 02°07′ N 53°42′	541/27	4073
24	Beeston Castle Cheshire	109	W 02°42′ N 53°07′30″	CPE UK 1935	1185
25	Old Sodbury Gloucestershire	156	W 02°21′ N 51°31′30″	106G UK 1416	3218
26	Coleby near Lincoln	113	W 00°32′ N 53°08′	541/111	3038 3039 4038 4039
27	Lealholme near Whitby	86	W 00°49′ N 54°28′	106G UK 1700	2252
28	Hazeley Heath near Reading	169	W 00°56′ N 51°18′30″	106G UK 1647	6018
29	Arthur's Seat Edinburgh	Scotland 74	W 03°09′ N 55°56′30″	106G Scot UK 119 pt I	5055
30	Ardmore Bay Mull	Scotland 53	W 06°08′ N 56°39′	CPE Scot UK 275	3119
31	Water of Fleet South Scotland	Scotland 92	W 04°16′ N 54°58′30″	106G Scot UK 52	3062
32	Kirkdale Burn South Scotland	Scotland 92	W 04°18′30″ N 54°52′	106G Scot UK 42	3069 3070

Plate 17

Plate 18

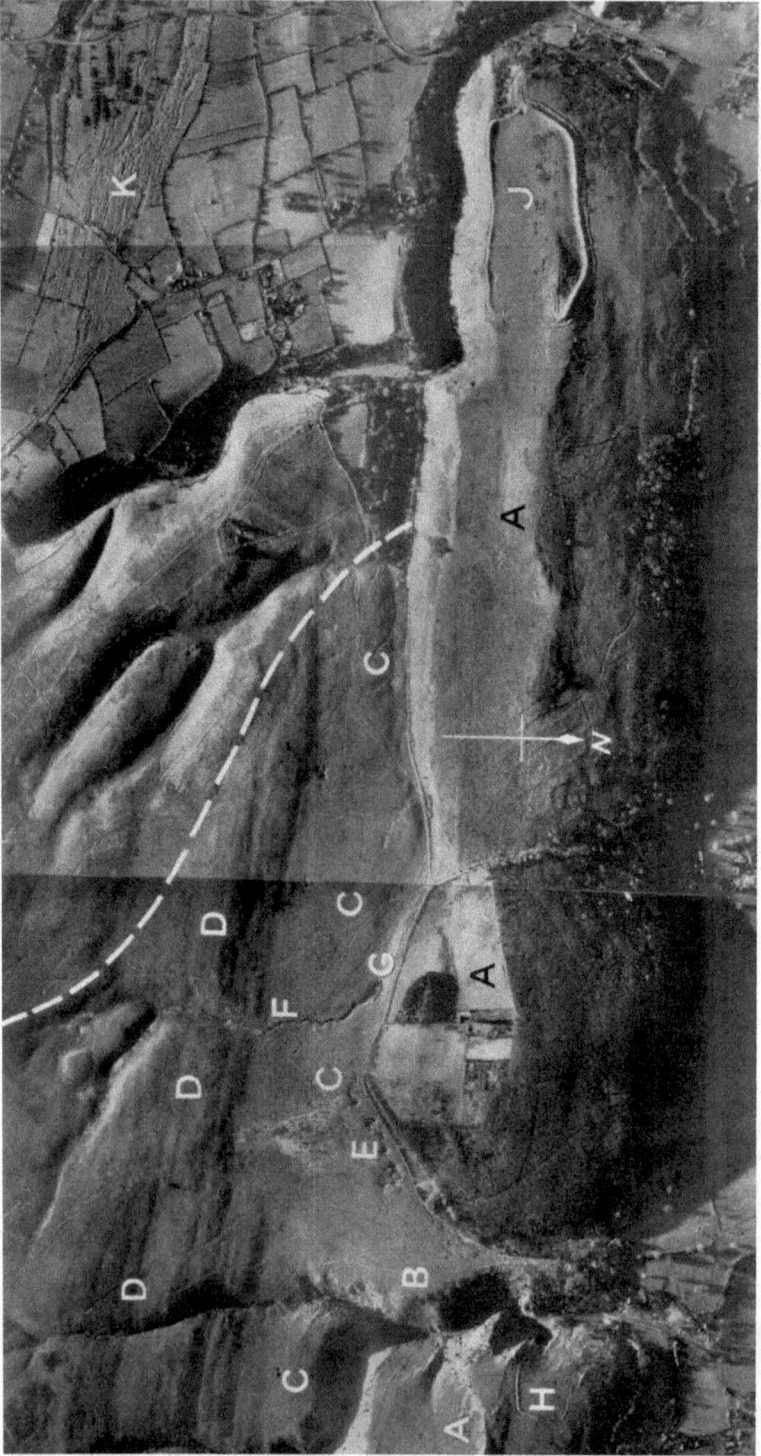

Plate 19

Plate 20

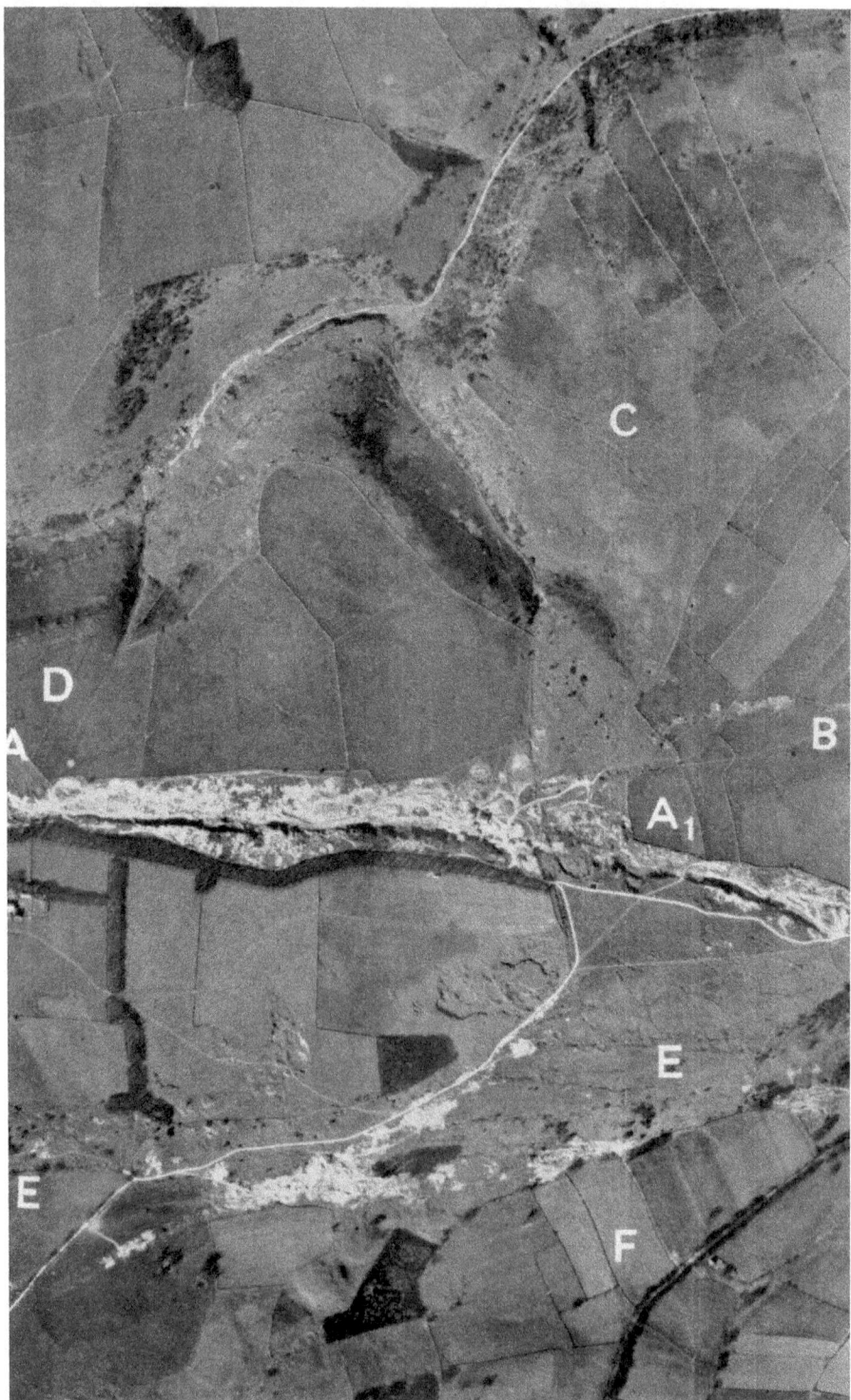

Plate 21

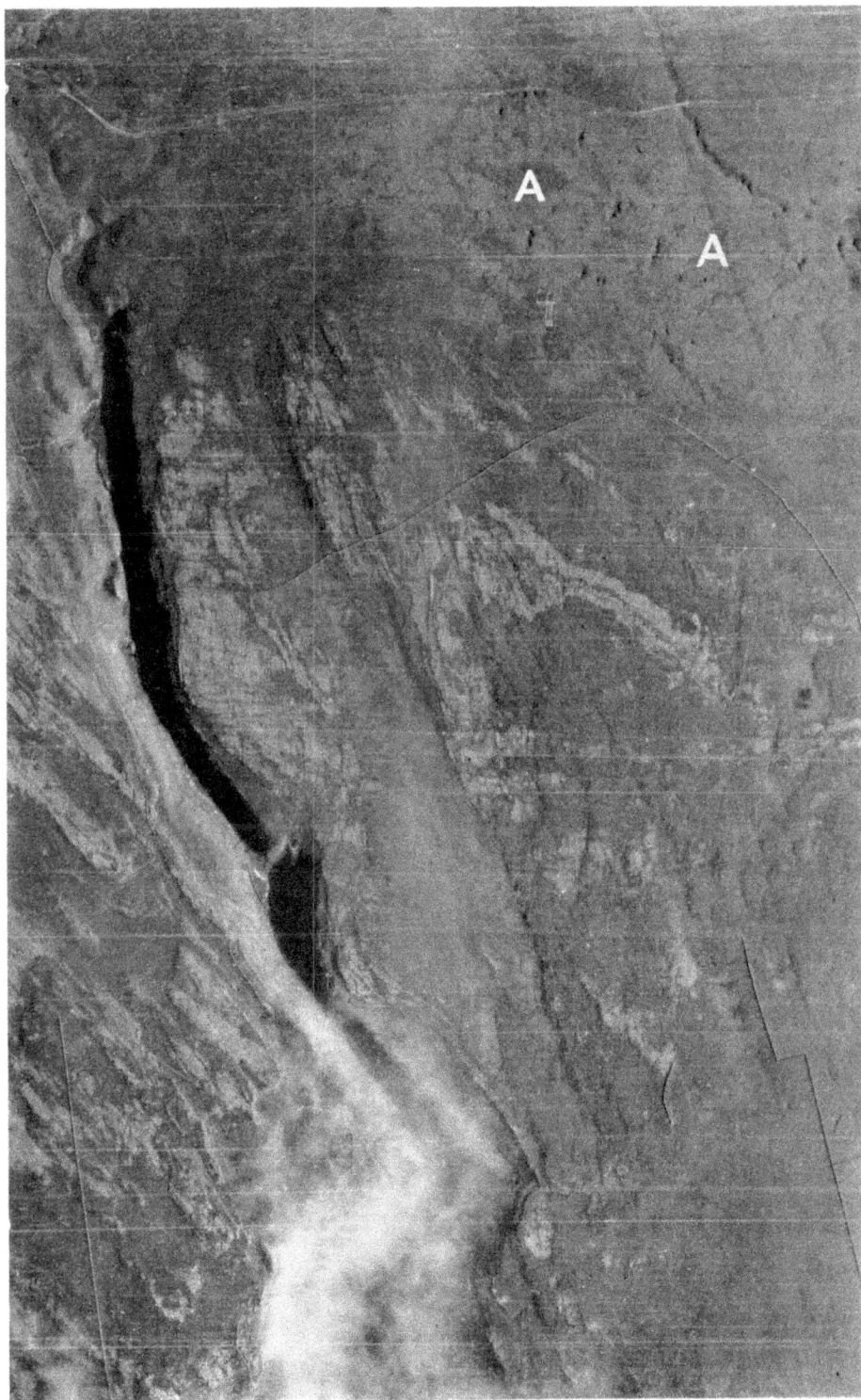

Plate 22

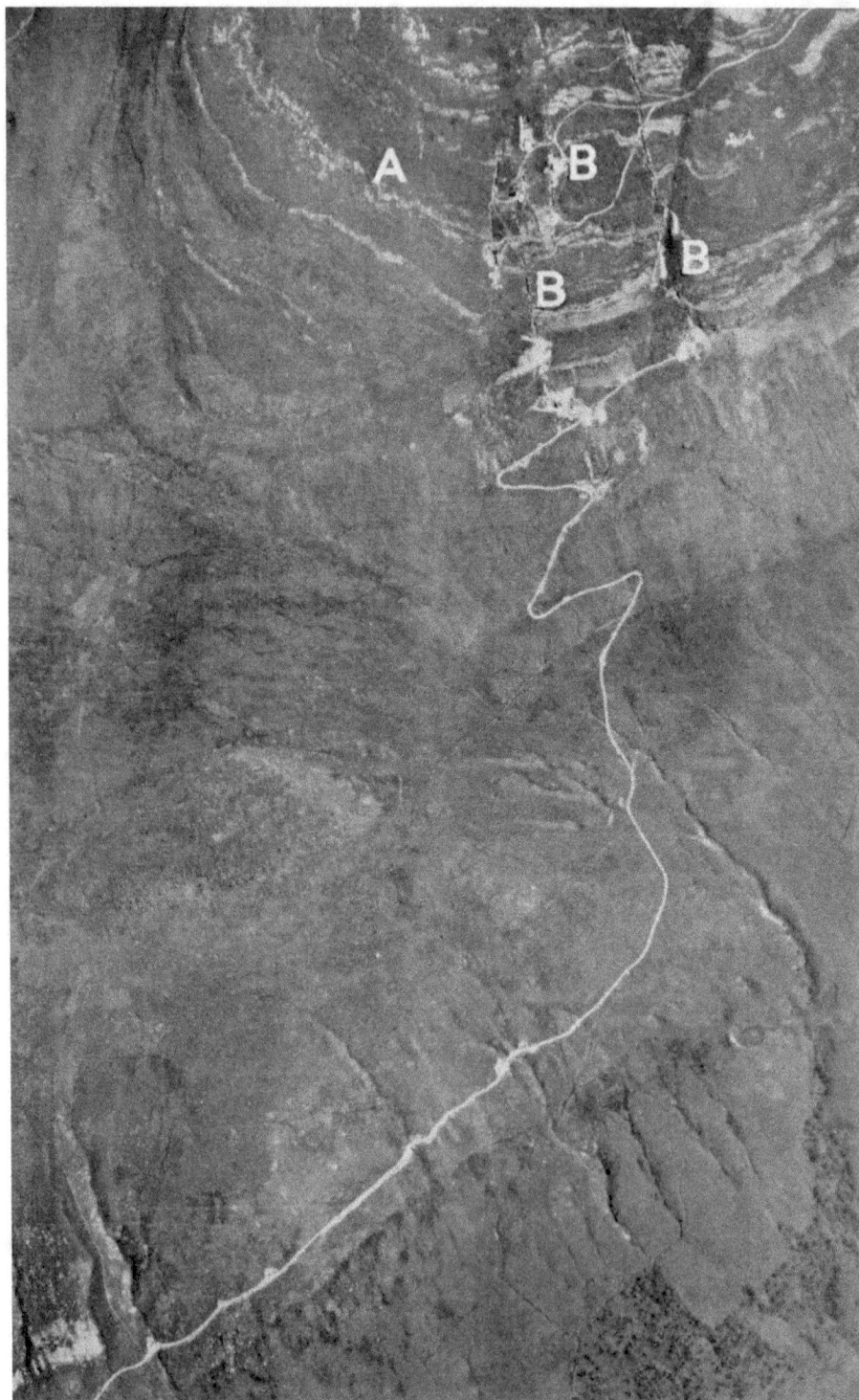

Plate 23

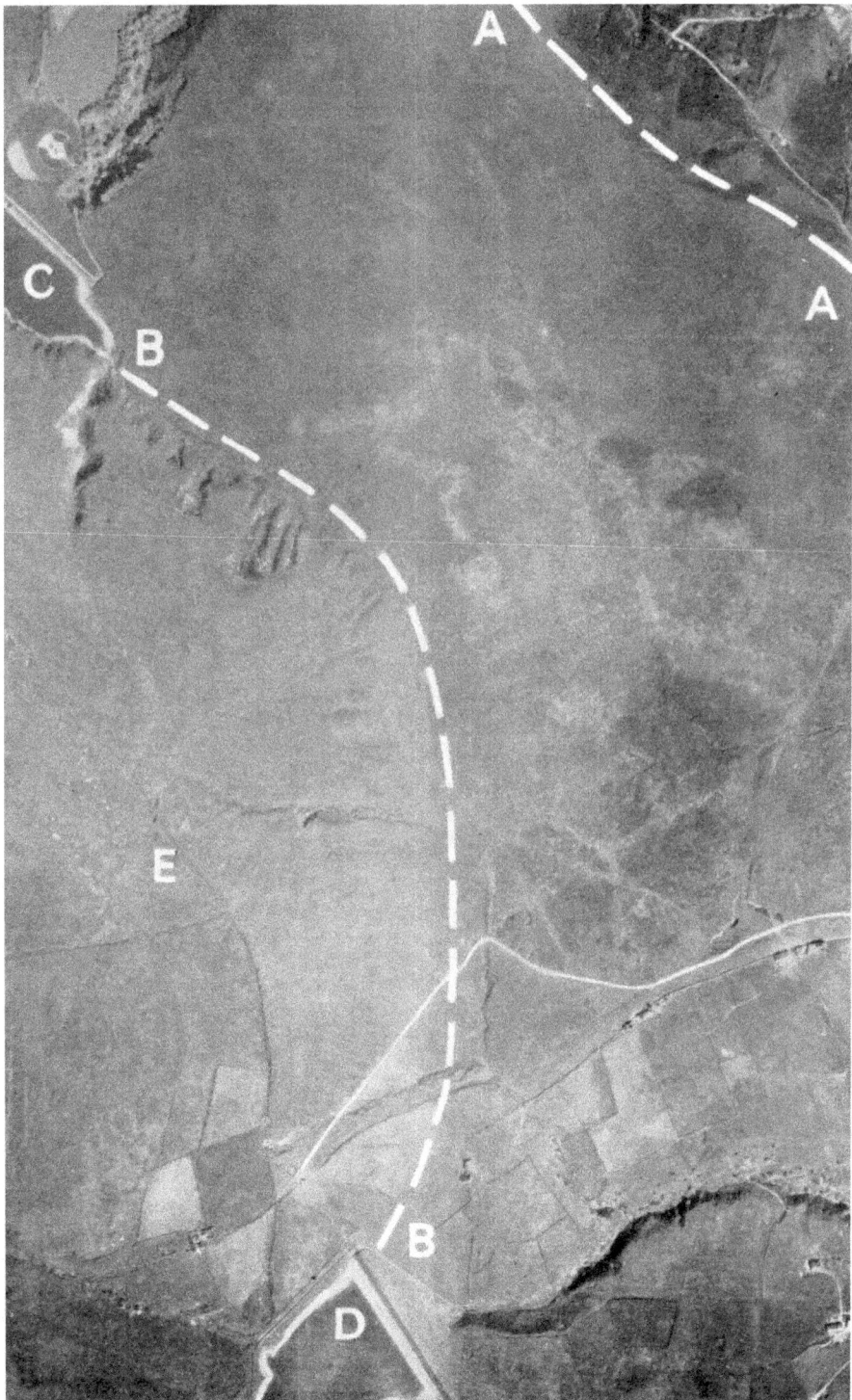

Plate 24

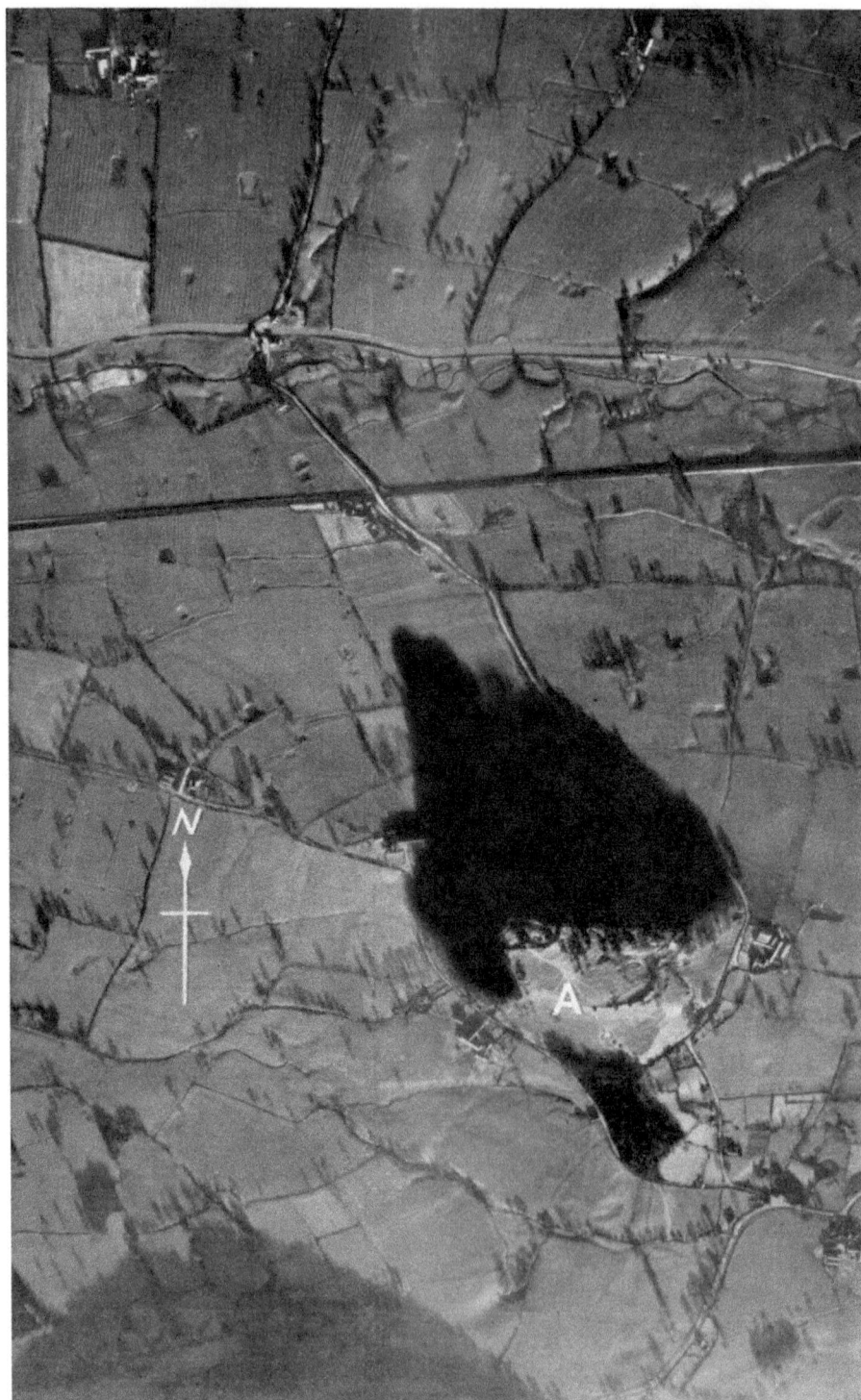

Plate 25

Plate 26

Plate 27

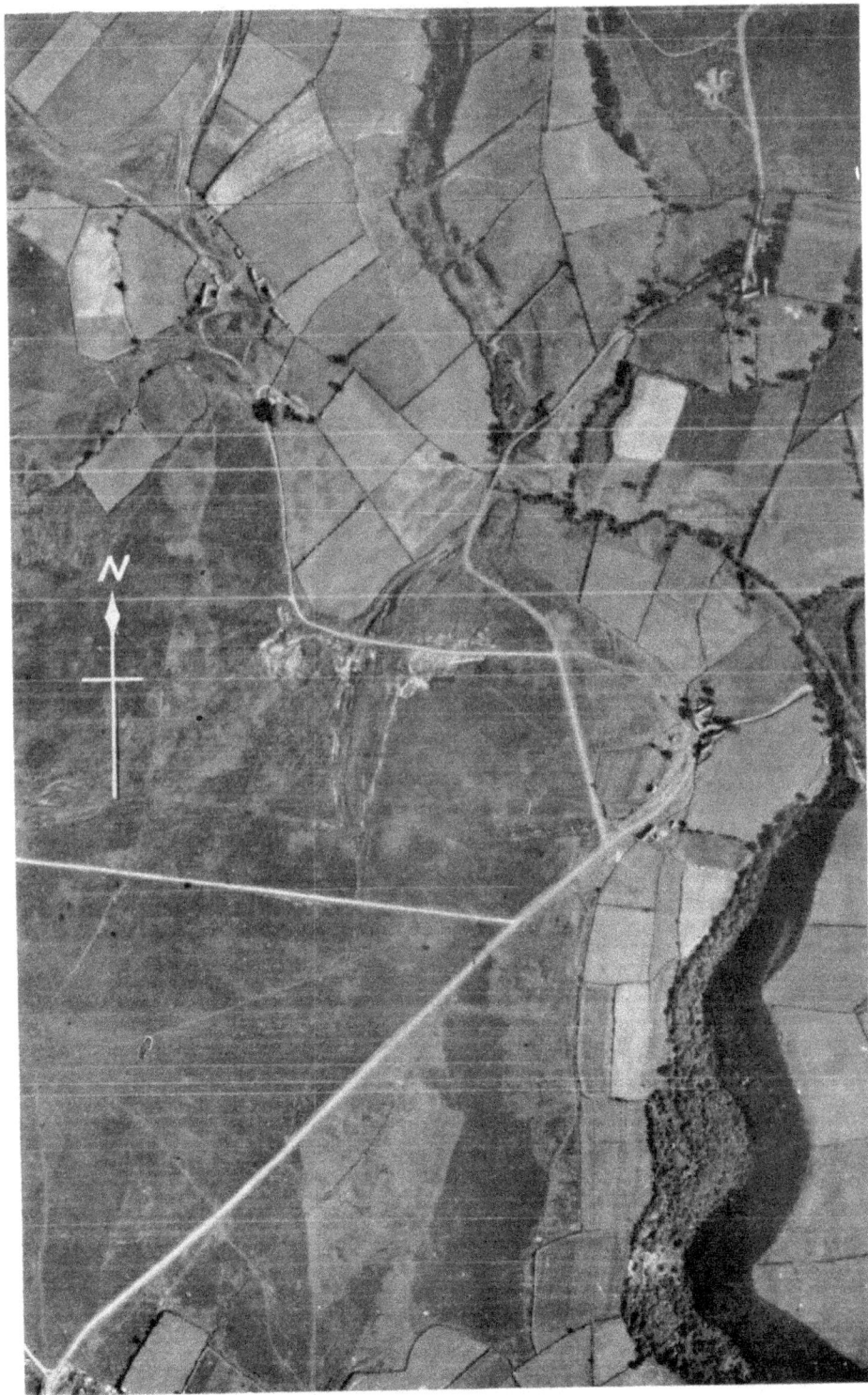

Plate 28

Plate 29

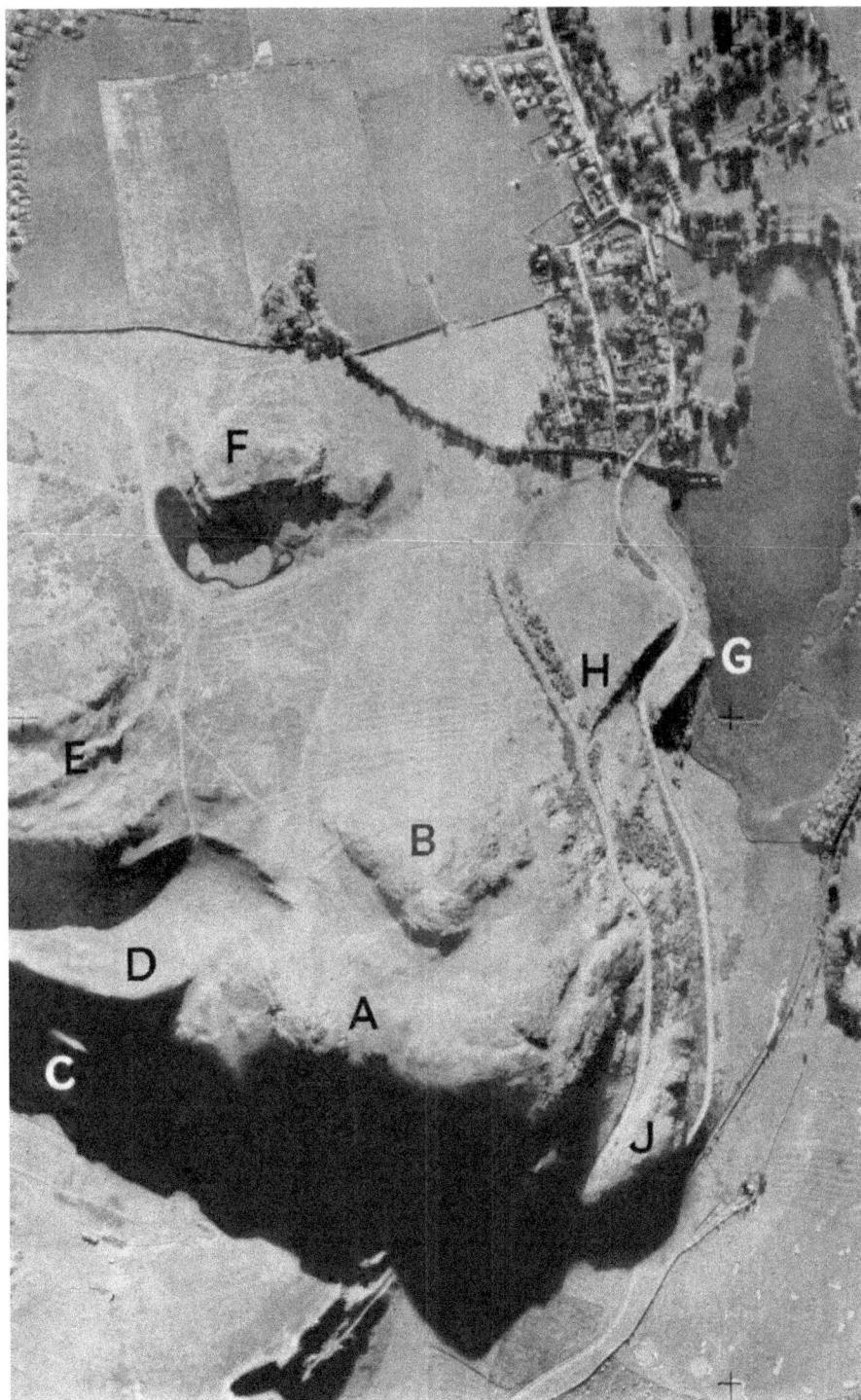

Plate 30

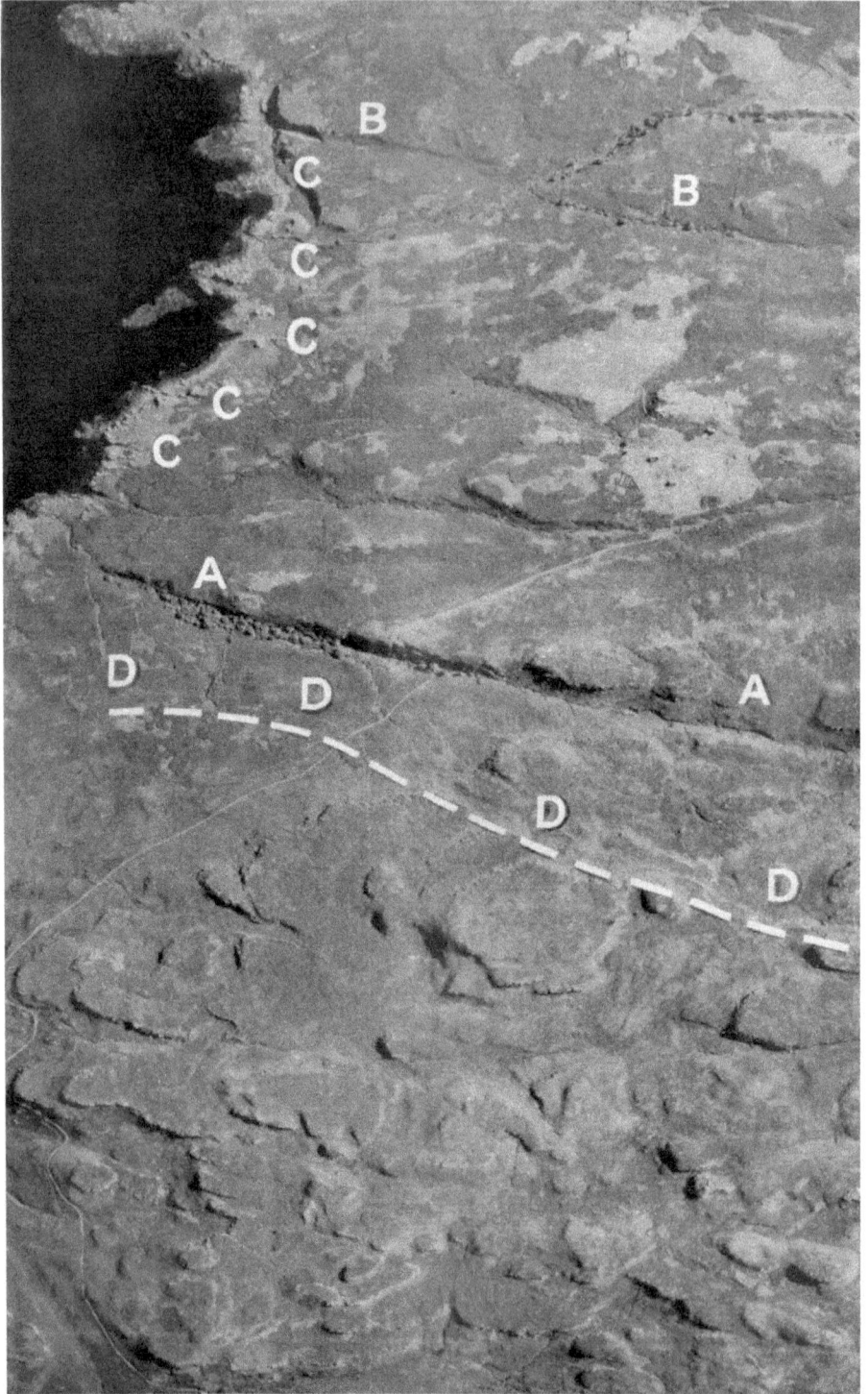

Plate 31

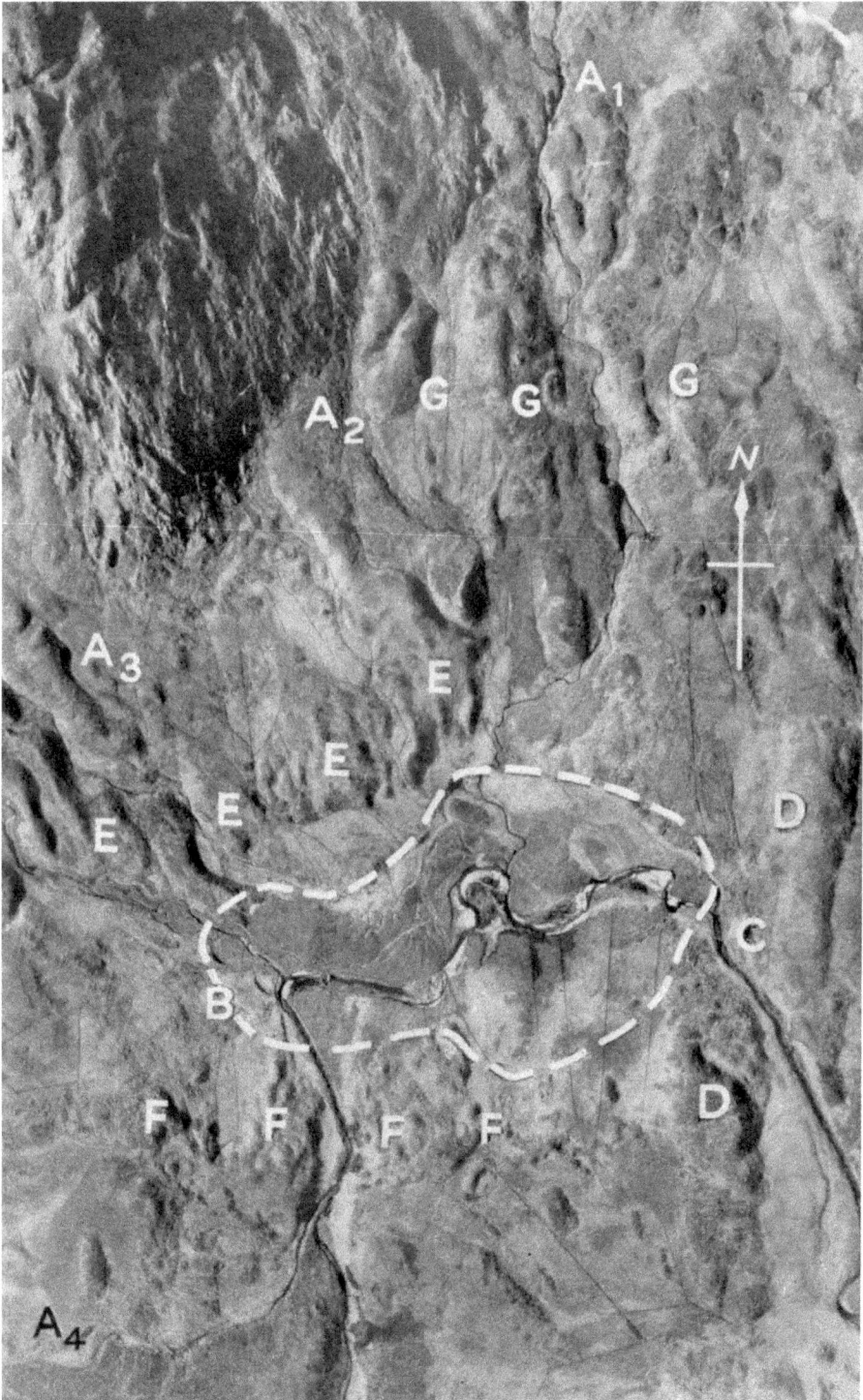

Plate 32

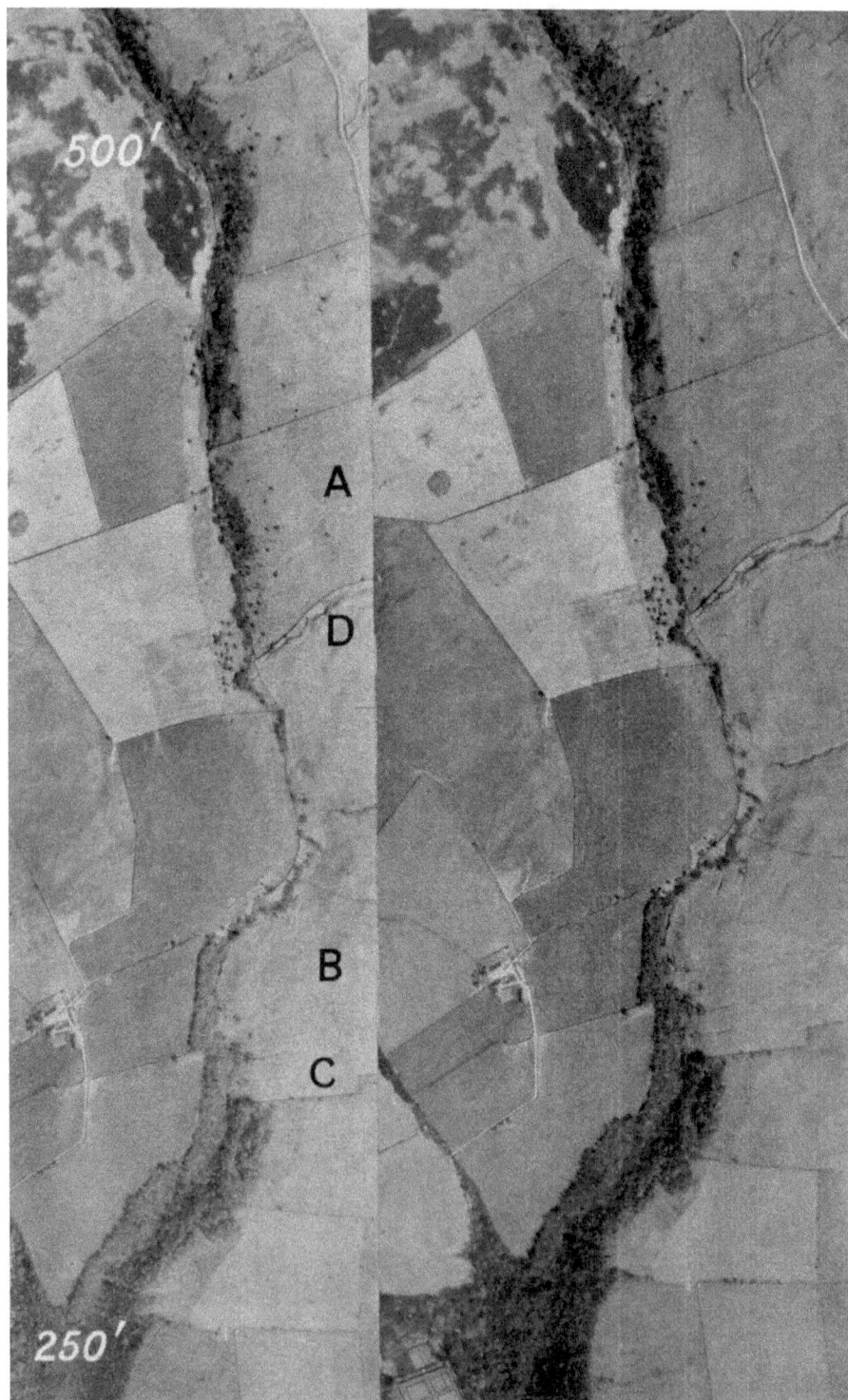

was probably an additional important factor influencing the siting of the villages along this junction. The change in landscape on the Lias is barely appreciable on the present photograph, though in the area West of Harmston and Coleby the patches of woodland and hedges suggest the approach to the less open landscape of the lowlands to the West.

Though the Jurassic rocks of the North York Moors are regarded as contemporary with those of the Midlands and Southern England, they were deposited under different and separate conditions and from a geographical point of view they produce an entirely different landscape, particularly in their Northern portion round the River Esk. Plate 27 is therefore designed to show this highly contrasting aspect of the Jurassic rocks and includes a portion of the moorland area known as Lealholm Rigg, lying immediately to the North of the Esk valley. The uncultivated area shown on the print is a region of rolling moorland with a very thin soil cover and a vegetation of heather and mosses and very poor grass. It coincides with the outcrop of the Lower Oolite which in this area consists of an estuarine series of Sands, Clays and Shales totally dissimilar from the Oolites of Southern England, and, as the photograph demonstrates, producing a totally different landscape. The whole of this region has been heavily glaciated, but on the Oolite moorlands there is comparatively little drift cover, and in the area of the present photograph the physical geography of the upland may be regarded as that arising from the nature of the estuarine beds themselves. The limit of the moorland to the North-east marks the line of a fault beyond which the Kellaway Beds appear, but the contrast along this line is probably due as much to drift conditions as to the solid geology. Here the Jurassic rocks are overlain by a considerable thickness of Glacial Sands and Gravels, and this drift cover extends Southwards on either side of the Stonegate Gill to the East of Lealholm Rigg. The thick and comparatively fertile soils of the drift, combined with the lower relief of this tract have therefore led to the highly contrasting landscape of a patchwork of enclosed fields which is characteristic of the valleys and dales within the North York Moors and which serves to re-emphasize the bleakness of the Moors themselves. Stonegate Gill, in common with most of the Northern tributaries of the Esk, appears to have had its origin as a stream following the Southward dip of the rocks, and it will be noticed that it has excavated a deep notch in the area shown in the South-eastern corner of the photograph where it approaches the deep glaciated valley of the Esk.

The study on air photographs of the more recent sedimentary formations presents rather greater difficulty, especially as they are largely confined to areas of lowland with gentle relief forms and with almost continuous cultivation which makes identification of boundaries difficult even on the ground. Plate 28 shows an area about 10 miles South-south-east of Reading where it is possible, however, to identify on the photo-graph the distinction between two sections of the Eocene, the London Clay and the Bagshot Beds, though it must be pointed out that in this case the limit of applicability of study on air photographs is being approached. The London Clay occupies the area marked *AAA* on the photograph and forms a shallow lowland which joins the valley of the Whitewater to the North-west. On either side the Bagshot Beds, marked *BBB*, form areas of rather higher land, though the slopes involved are very slight indeed and are barely perceptible on photographs without careful stereoscopic examination. The colour distinction between the generally dark-coloured London Clays and the yellowish sandy materials in the Bagshot Beds is therefore the main reason for the slight differences of tone on the photo-graph which make possible the recognition of the division, though the white appearance on air photographs of recently cultivated soils of all colours obscures the boundary in many cases. Differences of vegetation and land use between the London Clay and Bagshot Beds, which are recognizable in some parts of the London Basin are barely evident in this district though it is perhaps proper to point out that if park areas are ignored one can just appreciate the slightly greater proportion of grass-land and the wider distribution of trees on the Clay.

The area *C* in the North-east is a small portion of Hazeley Heath which lies on the Plateau Gravels (of doubtful age but probably of river-borne origin derived from the Chalk country to the South[1]), which overlie the Bagshot Beds. Here the distinction is sharply marked, and it should be mentioned that, in contrast with the less obvious divisions within the Eocene, wherever the overlying Sands and Gravels have led to the forma-tion of a heath landscape the recognition of such features on air photo-graphs presents no serious difficulties because of the highly distinctive appearance of heath country on the prints.

The photographs so far included to illustrate geological phenomena have been chosen to cover some of the major rock types occurring in the British Isles, and in references to later plates it will be possible to indicate the main characteristics of some of the other series not so far mentioned. Plates 29 and 30, however, are intended to serve as examples of cases where even minor geological features can be clearly identified on air photographs despite the fact that they cover only very limited areas and have a purely local geographical significance.

[1] See *Geology of the Country Round Aldershot and Guildford* (H.M.S.O., 1929), pp. 120, 122.

Plate 29 is chosen as a case where the detailed geological structure of a mainly intrusive igneous area can be seen to have a very direct relationship with the present sharply defined relief features. The photograph shows the greater part of Arthur's Seat near Edinburgh which, as a whole, may be described as a volcanic neck, but, for the present purpose, the main significance of the print lies in the fact that it is possible to identify the location of the Basalts which form the upstanding points among the softer materials of the neck. These have been marked A to J on the print, A and B being the two vents of Lion's Head and Lion's Haunch respectively. The edges of contemporaneous Basic Lava sheets tend to stand up as ridges above the surrounding areas of Tuffs, one example being the well-marked ridges at E, though perhaps the clearest case is that of the edge of the Dunsapie-type Basic Lava which forms the sharp lines of Long Row (D) and Loch Craig (H). Lion's Head (A), Lion's Haunch (B) and Dunsapie (F), although varying in detail, all include intrusive Basalts which stand up in sharp contrast with the surrounding Basic Agglomerate as does the great intrusion of Dunsapie-type Basalt in Samson's Ribs (J). The two ridges just visible in the shadow at C mark the Southern end of the Dasses which are sills of Dunsapie-type porphyritic Basalt, and at G the Girnal Craig marks the line of the same sill to the South of the main volcanic neck.

The lowlands surrounding Arthur's Seat are uniformly covered with a deposit of Boulder Clay, though there is a certain amount of alluvium in the immediate vicinity of the lake at Duddingstone.

Plate 30 shows a portion of the Northern part of the island of Mull near Ardmore Bay where the absence or extreme thinness of soil or drift makes it possible to study the effects of even the most minor geological features. The main rock of the area consists of extrusive Basalt, and in this district, particularly in the part shown on the bottom half of the print, there has developed a remarkably fine example of trap featuring. This phenomenon, thought to be occasioned by the method of extrusion and cooling of the various Basalt sheets, produces what is now more commonly termed terraced Basalts, or formerly traps, which give to the landscape the extremely distinctive form of irregularly shaped platforms separated by the characteristic terrace-like steps. On a topographical map with a contour interval of 50 feet most of the smaller details of such a landscape inevitably disappear except for the generally irregular shape of the contour lines and the occasional presence of small 'closed' contours corresponding to individual upstanding sheets of Basalt. On a high-quality air photograph of the type shown, however, every detail of the terraces can be seen and the general effect is one which makes it extremely easy to envisage the 'flow' of material which was responsible for the present land forms.

The whole island of Mull is crossed by an extremely complicated network of igneous intrusions into the main extrusive mass, most of these taking the form of dykes usually with a width of some 5 or 6 feet. A great many of these dykes are of acid material and the distinction between the line of the dykes and the surrounding rock results in their forming recognizable surface features though none of them is of great dimensions. On Plate 30 the line of two of these acid dykes can be traced at *AA* and *BB* partly because, in both cases, the direction of the intrusion is followed by the course of a small valley. On the coast at *CCCC*, however, a wave-cut platform has been developed round Ardmore Bay and differential erosion has resulted in the actual material of the dykes themselves being left upstanding as wall-like ridges which can be seen quite clearly on the photograph despite their extreme narrowness. In this locality basic dykes are much less common and where they do exist they do not in all cases contrast with the surrounding rock so sharply as those composed of acid material. Nevertheless, it is just possible on the original print for Plate 30 to identify the line of the only basic dyke in the area *DDD*, though in this case it was thought advisable to draw a dotted line parallel with the dyke to facilitate recognition of this feature. The line of this dyke is most easily recognized by the lighter tone of the intrusive rocks themselves because, although its course is followed by a slight surface depression, this is significantly less well-defined than the valleys along the acid dykes.

Chapter 3

THE STUDY OF EROSION ON AIR PHOTOGRAPHS

A. *River Erosion*

The study of river erosion may be carried out with suitable air photographs to a degree of precision adequate for most geographical and geomorphological purposes. Provided that photographs of the correct type are available, modern photogrammetric equipment[1] can be used to measure changes of height ranging from 1 : 500 to 1 : 1,000 or even smaller proportions of the flying height so that it is clearly possible to produce sections and profiles of streams and valleys as well as plans of their course which will compare reasonably favourably with the normal work produced by means of ground survey. Such studies, however, call for moderately elaborate equipment and often for special photography, so that in the present section reference will be made rather to the identification of different forms and stages of river erosion and development which can be achieved with the aid of single photographs or preferably with readily available pairs of photographs and simple hand stereoscopes.

Plate 31 is intended primarily to show youthful streams in their torrent stage in a mountainous area, though the actual choice of print was partly governed by the fact that it also shows a number of other points of interest. The area lies in the Southern Uplands of Scotland in an amphitheatre formed by the Eastern slopes of the Granite mass of Cairnsmore of Fleet and the adjacent Craigwhinnie to the North, from which the Carrouch, Mid and Craiglowrie Burns flow Eastwards and Southwards to unite in the centre of the photograph as the Big Water of Fleet. The youthful headstreams $A_1A_2A_3A_4$ are here in their torrent stage with obviously ungraded courses, and it is possible in some cases, even in these very small valleys, to see that the stream bed is rock-strewn. Even on a single print it is possible to see, especially at A_1 and A_4, that the valley is little more than a V-shaped notch and in places quite a deep, narrow gulley, but of course this feature is much more readily appreciated when prints are studied under the stereoscope.

Before the remainder of the river courses on Plate 31 are discussed, it is necessary to explain the effects of glaciation upon this amphitheatre of hills. Local glaciers in the valleys on the Eastern side of Cairnsmore

[1] See below, Appendix.

37

and the Southern side of Craigwhinnie moved down into the area shown on the photograph and deposited a most impressive series of concentric moraines,[1] some of which have been marked as *DD, EEEE, FFF, GGG,* on Plate 31. In particular it will be noticed that the moraine *DD* lies across the line of the Big Water of Fleet, and before the cutting of the present deeper portion of the valley through this moraine below *C* this mound of glacial material apparently resulted in the section of the stream immediately upstream from *C* becoming graded and developing an alluvial flood plain on a small scale indicated by the dotted line. Across this level tract of ill-drained land between *B* and *C* the stream meanders in a fashion more characteristic of a river in a much later stage of development, the white patches on the photograph indicating areas of deposition of sandy material on the inner edges of the meander curves. With the progress of erosion of the stream through the moraine below *C*, however, the effective base line of erosion of the stream above *C* has been lowered, and careful examination will reveal that the meandering portion of its course is now being slightly incised.

The gradual widening and opening of the valley section, the progressive grading of the stream profile and the disappearance of some of its irregularities of direction can best be appreciated on air photographs by stereoscopic examination, and therefore Plate 32 has been arranged as a stereopair which may be examined with a hand stereoscope though a general impression can be obtained without its use. From the marked heights it will be seen that, assuming a photographic scale of 1:10,000, the gradient in this portion of the Kirkdale Burn is about 1 in 20, indicating that from the point of view of profile it is still comparatively youthful. Nevertheless, stereoscopic examination reveals that, unlike the upper portions of the streams shown on Plate 31, the Kirkdale Burn has no appreciable irregularities of profile and has in fact a remarkably uniform gradient throughout its length. The course of the stream in this photograph has marked curves but in detail it does not show the extreme angularity of changes of direction characteristic of the mountain tract. It is, however, the valley section which is of greatest interest. It will be seen that the valley is still distinctly V-shaped in section, but here the notch is being widened appreciably and near points *A, B* and *C* is approaching a more rounded stage of development. At these points, too, one of the mechanisms by which the process of valley widening is achieved can be appreciated, since under the stereoscope the lateral marks on the photograph can be recognized as rainwater gullies marking the action of local surface

[1] See Charlesworth, *Trans. Royal Soc. of Edinburgh,* vol. LV, pt. I, no. I, and the *Memoir of the Geological Survey,* Sheet 4, p. 22.

water in wearing away the steep sides of the valley. At *D*, moreover, a further stage in this process is to be observed in the development of a tributary valley producing a major break in the valley side of the main stream.

Examples of the following stages of increasing maturity of river valleys are too numerous to permit of the selection of an entirely typical example, especially as the exact character of the valley form and river development in this stage is often determined as much by local geological conditions as by the theoretical erosional phenomena. It is therefore suggested that reference be made to Plates 14, 19, 49, 67, etc., as examples of river and valley forms developed under widely varying circumstances.

The later stage of river development associated with the formation of flood plains and the other phenomena of well-graded streams is particularly susceptible of study on air photographs since colour differences allow the areas of deposition to be detected, and differences of texture and moisture content make it possible to trace the progress of this particular phase in the cycle of erosion and deposition.

Plate 33 represents a slightly abnormal case in that the lowland shown on the photograph is not a simple river valley but part of the coastal lowland to the North of Luce Bay in South-west Scotland, where river alluvium, raised beach material and blown sand are intermixed. Nevertheless, it is a remarkably good example of some of the mechanisms of a meandering stream and illustrates the ease with which these can be identified from the air. The main area of alluvium is separated from the present beach by a very extensive area of blown sand (*CCC*) which invariably produces a highly characteristic appearance on the photographs and of which further examples are included in the section on coastal development. At *D*, an isolated area of blown sand has been incorporated in the agricultural area of the surrounding alluvium, but the 'blotchy' appearance of the field in question makes the distribution of this material amply evident.

The elaborate and complicated pattern of meanders of the Piltanton Burn require no special comment, though perhaps it should be pointed out that here the meandering probably developed on an already low-lying and nearly horizontal plain so that the photograph contains no good evidence of the way in which meanders may produce such a level tract by the processes of lateral and downstream erosion. At *E* there is a remarkably good example of a cut-off meander which is rendered all the more distinct because of the black appearance of the still water in the small lake—a phenomenon explained above (see p. 7). At *F*, *G*, *H* and *J* there are four other examples of cut-off sections of the stream, though these are now dry or partially so and do not have the same sharpness of outline as E. At K there is a particularly interesting case of an elongated cut-off meander which is only traceable in most of its course as a shadowy mark across the cultivated fields but which is nevertheless much more readily observed on the air photograph than on the ground. At the points marked L on the plate there are evidences of former courses of the river in the form of curved outlines of darker colour representing the limits of the uncultivated ground on the less well-drained material of the stream bed, whilst at M the shadowy appearance arises from the presence of a former ill-drained area round the tributary O which is now largely hidden on the ground by the progress of cultivation.

Plate 34 shows the valley of the Water of Fleet to the North of Gatehouse of Fleet in South Scotland, where conditions approach more nearly to the theoretical conception of a well-graded river working over the floor of its valley and gradually producing a flood plain comparable with that shown on Plate 33. The present stream occupies the centre of the plain, but evidence of earlier positions is to be seen in traces of former meanders at B, B', B^2, B^3 and B^4. On this photograph the white patches A to A_6 are areas of sand and shingle which are being deposited on the inner banks of the meander curves, whilst the effects of the corresponding erosion on the outer edges of the curves can be seen at A', where the steep outer bank indicates the fact that this particular meander is being cut into slightly higher ground, and at B^4 where a curved terrace-like cut in the field marks the limit of erosion of the outer edge of a former meander.

At A and A' shrubs and small trees can be seen to have occupied portions of the recently deposited sand and shingle and of the sloping portions of the immediate banks, and throughout the course of the river on this photograph practically all the tree growth is confined to similar localities. The main areas of alluvium are almost entirely devoid of trees in contrast with the surrounding districts of which only a small portion can be seen, and in the main the soils of the flood plain are devoted to arable agriculture. The two most notable exceptions are at E and E_1 where there are areas of poor meadow, the former being an ill-drained area where a left-bank tributary occupies a number of alternative courses before joining the main stream. At E_1 the grass area lies on a second less well-drained portion of the plain near the former meander at B, and it is of interest to notice that the artificial cut along the field boundaries at F is being used to improve the drainage along the line of this former portion of the river course.

The two main roads of the area CC and DDD characteristically avoid the portions of the plain which have been liable to flood, and in general terms their course may be said to mark the limits of the alluvium of the flood plain. Though it is obvious that settlement would tend to be more or less along the line of the roads, it is nevertheless significant that there is not a single building on what is now the highly cultivated low-lying plain.

Plate 35 shows a portion of the valley of the Water of Ken at New Galloway, immediately to the North of Loch Ken. The presence of the lake just to the South of the area shown on the photograph has meant that the level of the water in it has acted as the base line of erosion for this portion of the river. Conditions have recently become entirely abnormal here since Loch Ken now acts as one of the storage reservoirs for the Galloway hydro-electric scheme so that the level of the water is subject to unnatural fluctuations owing to the fact that most of its intake and the whole of its outflow is controlled. Consequently, the area round BB and D on Plate 35 is subject to inundation by the waters of the loch and indeed is marked on the Ordnance Survey map as being part of the loch. Nevertheless, the periodic flooding has preserved traces of former conditions and it is therefore all the easier to follow the line BBB of a former course of the river, one part of which still retains a considerable volume of water. The Ken itself, therefore, appears to have shifted across this section of its valley and is now asymmetrically placed at the extreme edge of the flood plain. At C traces of a more recent course of the river can just be discerned, whilst at D there is a nearly complete example of a characteristically shaped cut-off meander lake.

The almost perfect grading of this portion of the Ken means that the relationship between its velocity, load and course is a critical one and even the slightest curve is liable to produce sufficient reduction in velocity on the inner bank to give rise to areas of deposition like those at $AA'A_2$.

The present relatively straight course of the Ken arises from the fact that it has now reached the Eastern limit of the plain and runs immediately under the steeper slopes of the straight-sided glacial valley along which the ice advanced Southwards, parallel with the river course, towards the rock basin of Loch Ken.[1]

It will be noticed that again the roads, both to the East and West of the river, keep to the higher ground as does the elongated village of New Galloway (GG). Though the whole circumstances are not apparent on the photograph it is perhaps worthy of note that the function of New Galloway was, in part, that of a river-crossing town on the so-called Old Edinburgh road (EEE), but that the road had to curve upstream about a

[1] See Charlesworth, op. cit., pp. 3, 11, 22.

mile to a point where the flood plain narrowed sufficiently to permit of bridging at Ken Bridge, though the latter site on the flood plain itself has significantly not developed into a sizeable settlement.

A minor point of some interest is the downstream diversion of the tributary confluence at F, a normal flood plain phenomenon accentuated in this case by the obvious presence of deposited material at A.

In the more normal circumstances under which sea level acts as the base line of erosion it is by no means uncommon for this base line to be raised or lowered in comparison with the surrounding land by changes in the relative levels of land and sea with consequent drowning of coastal reaches or, alternatively, renewed vertical erosion and the rejuvenation of the rivers. Examples of drowned coastal areas are included in the section on coastal forms and it is therefore only proposed here to include examples of the effects of rejuvenation.

Plate 36, since it includes a portion of the coast itself where changes of level are demonstrable, may serve as an initial example on a small scale of the process of rejuvenation at work. The area lies on the Western side of the estuary of the Nith in Carse Bay to the North of Kirkbean where raised beaches have been mapped in some detail. Practically all the land on the seaward side of the road (AA_1) consists of such a raised beach, caused by the apparent raising of the land level, the line of the road following the former coastline, except near the bottom of the print, where higher land can be seen on the right of the road. In consequence of the change the base line of the stream (B) was lowered after it had achieved a partly graded profile and developed a meandering course across the former beach. The result is, therefore, that further vertical erosion has led to the meanders becoming incised quite deeply into the material of the raised beach.

As this photograph was taken at low tide it also demonstrates the fact that for an appreciable part of each day the effective base level of erosion of the stream is the level of the low-water mark and therefore it has excavated a channel for itself across the present beach zone between the high- and low-water marks. Should a further rise take place in the height of the land relative to the sea, one may thus envisage a repetition of the events which produced the incised meanders upstream from C and the creation of a very similar new portion of the river below that point across a new raised beach.

The light, easily cultivated soils of the raised beach are now entirely concerned with arable agriculture, but their formerly ill-drained character and liability to inundation may be deduced from the pattern of old drainage channels at EE_1E_2, and it is significant that the farms themselves are all on the higher side of the road, their construction ante-dating the working of the land on the raised beach.

The white arrow, dot and bars at *D* compose a ground-control mark built to establish a point of known location to enable this and adjacent photographs to be used for accurate survey and mapping purposes.

The juxtaposition of Plates 36 and 37 leads from a miniature example of incised meanders to one of the most impressive examples of this land form in the British Isles, since Plate 37 shows a portion of the lower course of the Wye a few miles upstream from Chepstow. When it is remembered that the river Wye is still tidal at this point, the depth of the incision can be appreciated from the marking of a 350-foot spot height and the short sections of form lines at approximately 250, 150, 100 and 50 feet, whilst the scale of the meanders may be deduced from the fact that the river itself is about 100 yards wide near point *F*.

The downstream movement of meanders occasioned by the under-cutting of the downstream outer edge of the river curves is a well-known feature of regions of incised meanders, and in Plate 37 this is indicated by the steep cliffs on the Southern banks of the river at *B* and *D*. At *D* the slope of the cliff, the presence of woodland and the precise position of the camera combined to produce the circumstance that the actual river bank is not visible, and a remarkable visual impression of the process of under-cutting is created. The ultimate stage in this process would normally tend to be the 'breaking through' of the neck of land at *A*, leaving the meander as a cut-off portion of the river course, a stage of which there are examples at different depths of incision in the lower part of the Wye. In its present form the high, narrow neck of land between the meanders is a distinctive feature of the land forms of the Wye valley in the region between Chepstow and Tintern.

Lateral erosion, though perhaps less important than that in the down-stream direction, is, nevertheless, most impressive in this section of the Wye and has formed precipitous slopes at *F*, *G* and *H*, though on the photograph these are far more apparent in the cliffs at *F* than in the two other cases where woodland masks the steepest slopes.

Both downstream and lateral erosion combine to cause the development of valley 'flats' on the relatively gently sloping tongues of land between the meanders. Here the more gradual slopes, indicated by the form lines, for example near *E* and *J*, demonstrate the progress of vertical erosion whilst the horizontal displacement of the meanders was taking place (a process easily visualized at *E*), so that it should be appreciated that the term 'flat' is singularly inappropriate except perhaps below the 50-foot contour. In this portion of the lower Wye the river itself has clearly almost reached its base line of erosion so that the balance between erosion and deposition is critical, and below the 50-foot contour recently deposited material may easily be eroded by any change in tidal conditions or river volume—a fact which is indicated by the obvious instability of the banks at *K*, *L*, *M*, etc.

The presence of high, narrow necks of land, partly surrounded by a broad river, particularly in Limestone country, has meant that the 'spurs' in the lower Wye are the site of a considerable number of earthworks indicating man's early interest in these localities for defence and settlement. At *C* and *C*₁ there are two camps which occupy ideal defensive sites, the former being clearly visible, though the presence of woodland makes the latter very difficult to identify.

It is again characteristic of an area with a very deeply incised river valley that practically all present settlements and communications are concentrated on the upland away from the precipitous and largely inaccessible river banks, though the gentler slopes of the so-called 'flats' are cultivated, in each case from a single farm located at the edge of the high ground. The only exception is made by the railway at *H*, in which case consideration of gradients has led to the use of a valley-side position and tunnels not unlike the situation seen in the Monsal Dale area (Plate 19).

In most cases, however, where streams have undergone rejuvenation the resultant renewal of vertical erosion produces land forms of a far less spectacular type than those associated with large-scale, deeply incised meanders. These commonly take the form of river terraces, often of very small dimensions, left behind as the stream excavated deeper channels into its pre-existing valley floor and slight breaks of slope in the long profile of the stream produced by backward cutting in the lower portions of its course. Under these circumstances a proper impression can only be obtained from air photographs by stereoscopic examination, and Plate 38 has therefore been arranged as a pair for viewing through a hand stereoscope. It shows the valleys of tributaries of the Lednock River (itself a tributary of the Earn near Comrie, North-west of Crieff) which have undergone several stages of rejuvenation associated with the rejuvenation of the Earn and Lednock and the cutting of the gorge of Deil's Cauldron (*A*). The largest terrace is that marked *B* at a level which appears to be important throughout the district, but at *C* and *D* lower and smaller terraces can be readily identified. A great many other small areas of terrace can be examined under the stereoscope but no identification marks have been placed on the prints as these tend to hide photographic detail and produce an irritating effect of movement when viewed stereoscopically. Most of the smaller tributaries in the area have marked breaks of slope associated with the stages of rejuvenation, one of which has been indicated near point E, though one or two others will also be apparent when the plate is examined stereoscopically.

Stream development

The development of stream patterns and river systems is normally only demonstrable by reference to areas larger than can be shown conveniently on a limited number of air photographs. This difficulty can be overcome by the use of large mosaics or of small-scale photography, but neither of these techniques lends itself to good-quality reproduction for book plates, and it has therefore been found necessary to restrict the present section to two examples of interesting and rather abnormal forms of river development which can be illustrated by the use of comparatively small mosaics.

Plate 39 shows a most remarkable example of river-capture on the Avon at Inchrory in Banffshire, about 11 miles North of Braemar.[1] The 'elbow' of capture is seen at B, where the present course of the Avon (AAA) turns sharply towards the North and North-west. In the area marked C the small stream which forms the present headwaters of the Don can be seen meandering over a very extensive area of alluvium on the floor of a wide and open valley which has the appearance of a considerable degree of maturity. It will be seen that the course of the Avon upstream from point B is directly in line with the present upper Don, and at D there is a very well-marked gap overlooking the portion of the Avon valley where it makes its abrupt change of direction. The Avon upstream from B is regarded as representing the original headwaters of the Don, flowing into the present Don basin through the gap at D, in which case the Don in this section of its course must have been a stream of considerable dimensions, well capable of excavating the wide valley at C.

The capture of the upper part of the Don by the Avon is attributable to the fact that below B the latter river is flowing generally parallel with the strike of the rocks and has been able to excavate its course comparatively rapidly along the limestone and soft black shale, whereas the Don was flowing across the structures and presumably eroding at a relatively slower rate. The degree of vertical erosion on the downstream section of the Avon may be appreciated from the considerable gorges at E, whilst the resultant rejuvenation of the former upper course of the Don has produced a central incised section in the Avon valley above B which

[1] See A. W. Gibb, 'On the relationship of the Don to the Avon at Inchrory in Banffshire', in *Trans. Edinburgh Geological Society*, vol. IX, p. 227; and A. Bremner, *Physical Geology of the Don Basin* (University Press, Aberdeen, 1931), p. 8.

is clearly visible under the stereoscope and can just be detected on the mosaic at *F*. The Builg Burn (*G*) was, of course, formerly a tributary of the Don and had graded its course to the level of the through Don valley before capture took place, with the result that it now meanders over an area of alluvium into which its present course is being incised as a result of rejuvenation caused by the rapid erosion of the Avon into which it now flows.

Plate 40 shows an example of river development where the problem of explaining a remarkable stream diversion has not yet been solved,[1] and it is included partly as an instance where good-quality air photographs might be used as the basis for future geomorphological investigations. The area is on the Cumberland coast between Ravenglass and Millom where the River Annas (*AAA*), after flowing almost due North–South parallel with the edge of the Lake District hills, turns South-west and West as it enters the area shown on the photograph, and then almost due South–North to enter the sea at *B*.

The direction of wave incidence on the coast, as shown at the time of photography, suggests that normal wave transference of beach material may explain the deflection of the stream in its coastal section, though such an explanation does not fit well with the known direction of movement to the North, and Steers expresses the view that no complete explanation of the coastal diversion is yet available.[1]

The higher diversion of the stream towards the North-west at *C* is also unexplained. The air photograph brings into prominence a virtually dry valley (*DDD*) (along the inland side of which the Millergill Beck (*EEE*) flows Northwards to join the Annas) and it seems reasonable to speculate whether or not investigation on the ground might reveal this valley as a possible earlier outlet of the Annas to the sea at *F*. It should perhaps be pointed out that here the surface is composed entirely of drift of a character and form which is complicated by the fact that the area lies along the junction of the Lake District and Irish Sea ice[2] so that the post-glacial evolution of drainage is a matter which still calls for considerable detailed study.

B. *Glacial Action*

Whatever interpretation is placed on the relative importance and rôle of erosion by ice and by water in glaciated regions it is convenient to group the land forms of such areas into those where erosional features

[1] See Steers, *Coastline of England and Wales*, p. 84.
[2] B. Smith, 'Glaciology of the Black Coombe district', in *Q.J.G.S.*, vol. 68 (1912), pp. 402–48.

are the more important and those where deposition represents the main geomorphological result of the glacial period. In the case of the former, at any rate in Britain, many of the best examples are in upland or mountainous country where the vertical view-point of the air photograph has obvious practical advantages. On the other hand, it is a characteristic of glaciated mountain areas that the relief forms tend to have abnormally steep slopes and on the air photographs these may cast dense shadows which hide important features in the valleys and on lower slopes. Accordingly it is essential to select photographs taken during a part of the year and at a time of day which results in the shadows being cast in a suitable direction. This has proved possible in the case of Plates 41 to 45, but it must be understood that the existence of deep shadows may render this type of study for some individual districts impossible without specially arranged photography.

In the case of the valley glaciers of Britain the upper parts of the area of ice movement are now represented by the corries of the mountain tracts above the valleys, and Plate 41 shows an example of this type of land form on the Eastern slopes of Pen-yr-Oleu-wen, to the North-east of Snowdon, known as Ffynnon Lloer. This corrie, together with Ffynnon Llugwy farther East, appear to have been the sources of relatively small glaciers which joined the ice moving down the valley of Afon Llugwy towards the main Conway valley. It demonstrates clearly the characteristic back wall of the corrie since the slope from A (3,200 feet) to the floor of the corrie at B (2,200 feet) is extremely steep throughout, the upper 450 feet in particular being virtually a cliff face. Post-glacial erosion and weathering has resulted in this wall of the corrie being worn backwards into the hill mass, a process rendered evident on the photograph by the white streaking on the better illuminated side of the corrie which represents the areas of downwash on the steep slopes. Moreover, the ledges in the face of the wall and the floor of the corrie near B are covered with masses of angular rock fragments detached from the slopes above by the normal processes of weathering. This recession of the corrie wall has resulted in the hill mass being reduced to a knife-edge ridge which, at A, is little more than 30 yards wide and beyond which the photograph shows a remarkably fine series of gullies. On this Western face there are no corries and the mountain side takes the form of a remarkably straight but extremely precipitous slope forming the Eastern side of the glaciated valley of Nant Ffrancon (see Plate 43) on which a very large number of small independent watercourses commence in the gullies already mentioned and provide fine examples of streams in a very youthful stage of development.

PLATES 33—48

Plate	Locality	Ordnance Survey Sheet no. 1:63,360	Latitude and Longitude	Sortie number	Print numbers
33	Piltanton Burn near Glenluce	Scotland 90	W 04°53' N 54°52'	106G Scot UK 42	4122
34	Water of Fleet South Scotland	Scotland 92	W 04°11'30" N 54°53'30"	106G Scot UK 42	4346
35	River Ken New Galloway	Scotland 87	W 04°08' N 55°04'30"	106G Scot UK 52	4356
36	Carse Bay Nith Estuary	Scotland 92	W 03°35' N 54°55'30"	106G Scot UK 42	4003
37	Lower Wye near Chepstow	155	W 02°40' N 51°40'	CPE UK 1997	4002 4003
38	River Lednock near Crieff	Scotland 63	W 04°00' N 56°24'30"	106G Scot UK 37	3036 3037
39	River Avon Inchrory, Banff	Scotland 44	W 03°21'30" N 57°09'30"	541/388 106G Scot UK 82	3447 3449 4025 4437
40	River Annas Cumberland	88	W 03°24' N 54°16'	106G UK 1127	3002 3019 4019
41	Ffynnon Lloer near Capel Curig	107	W 04°00' N 53°08'30"	CPE UK 2615	4113
42	Ben Cruachan near Oban	Scotland 61	W 05°08' N 56°25'30"	CPE Scot 327	3368
43	Nant Ffrancon North Wales	107	W 04°02' N 53°08'	CPE UK 1996	2337 3335
44	Near Dolgarrog North Wales	107	W 03°51' N 53°12'	CPE UK 1996	1276
45	Palnure Burn near Newton Stewart	Scotland 87	W 04°20'30" N 55°01'	106G Scot UK 52	3274
46	Near Castle Douglas	Scotland 88	W 03°55' N 55°03'	106G Scot UK 52	4234
47	Arvie Burn South Scotland	Scotland 87	W 04°03' N 55°02'	106G Scot UK 52	3247
48A	Drumabrennan near Newton Stewart	Scotland 87	W 04°40' N 54°58'	106G Scot UK 52	3097
48B	Knowe near Newton Stewart	Scotland 87	W 04°38' N 55°00'	106G Scot UK 52	4302

Plate 33

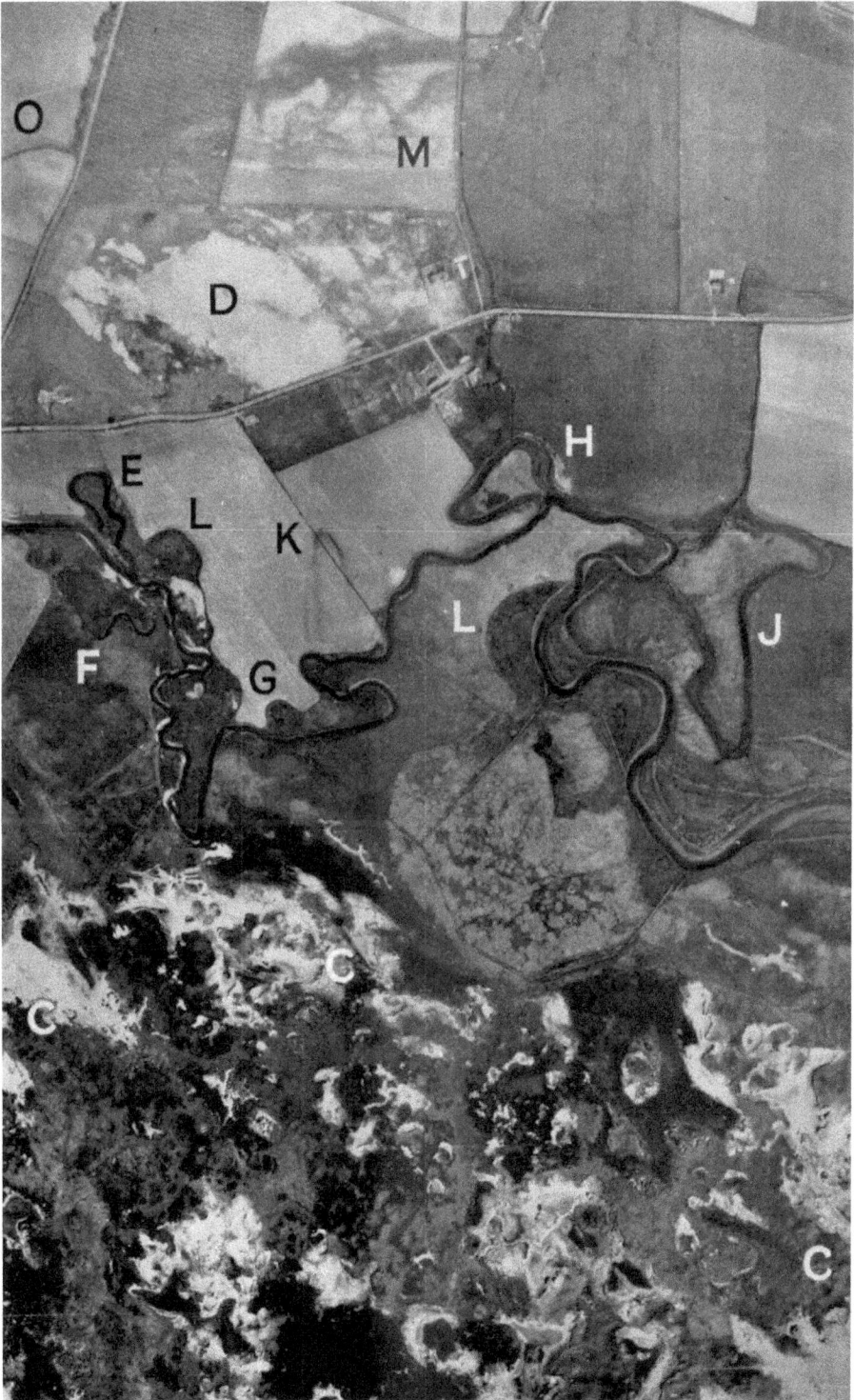

Plate 34

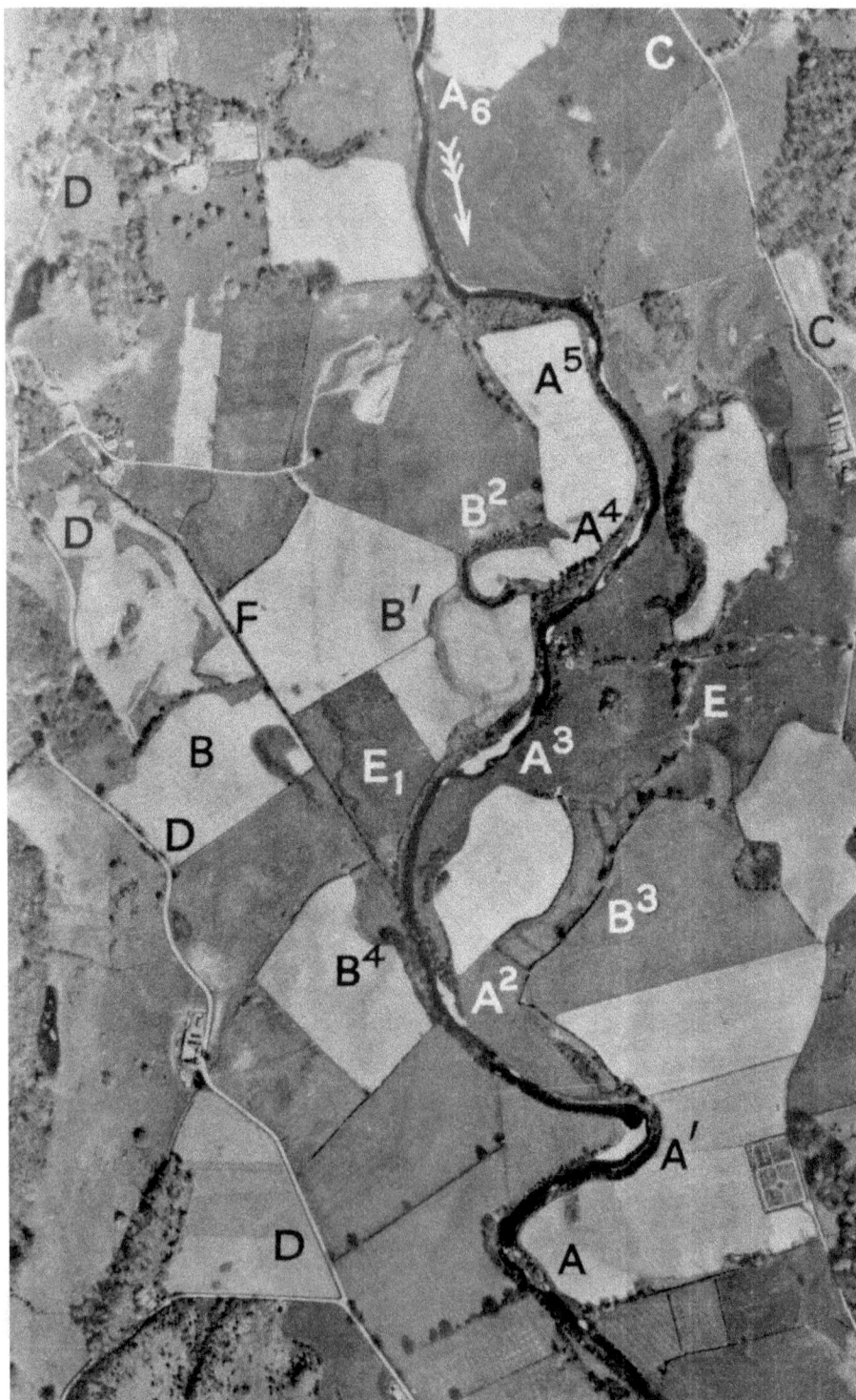

Plate 35

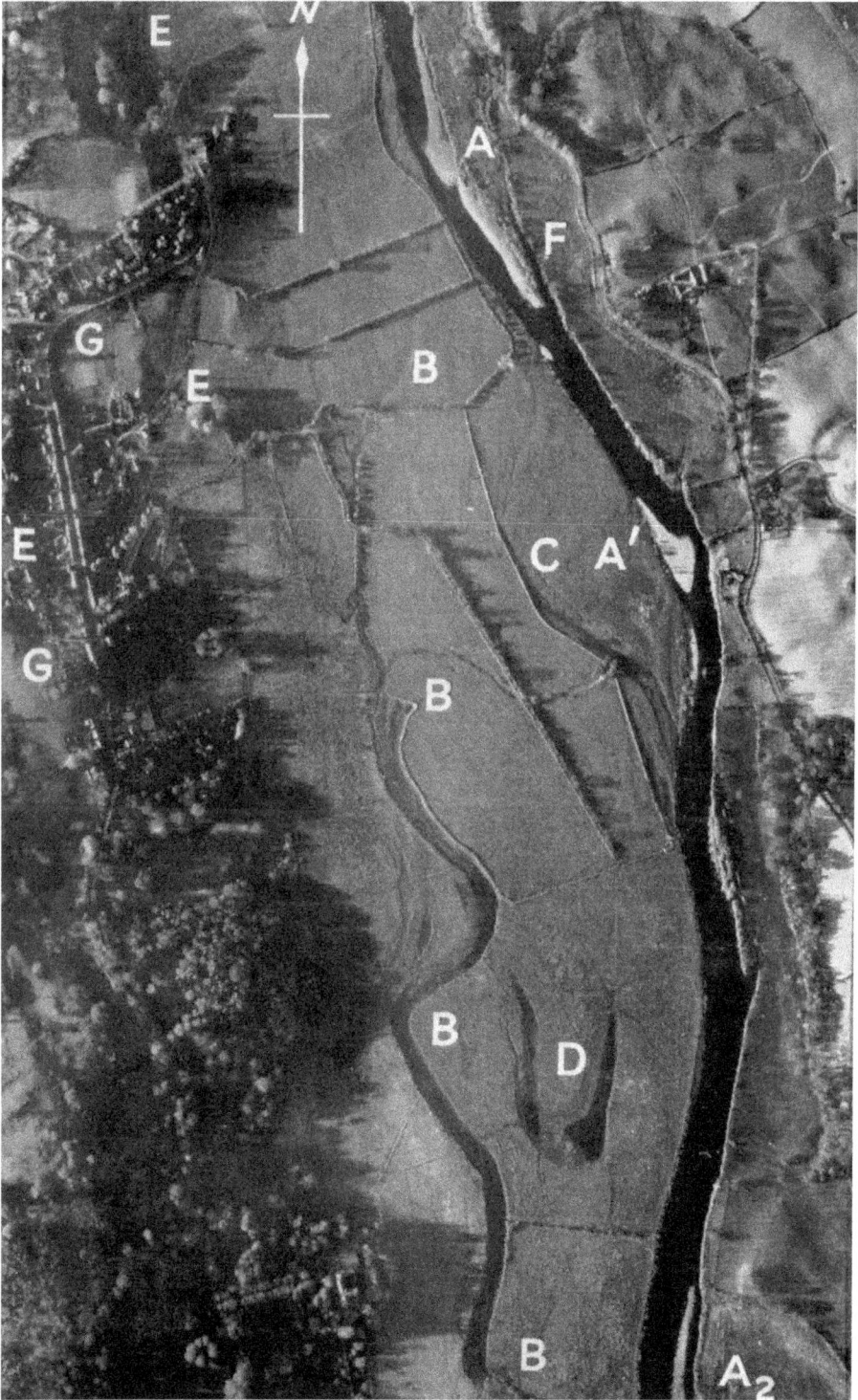

Plate 36

Plate 37

Plate 38

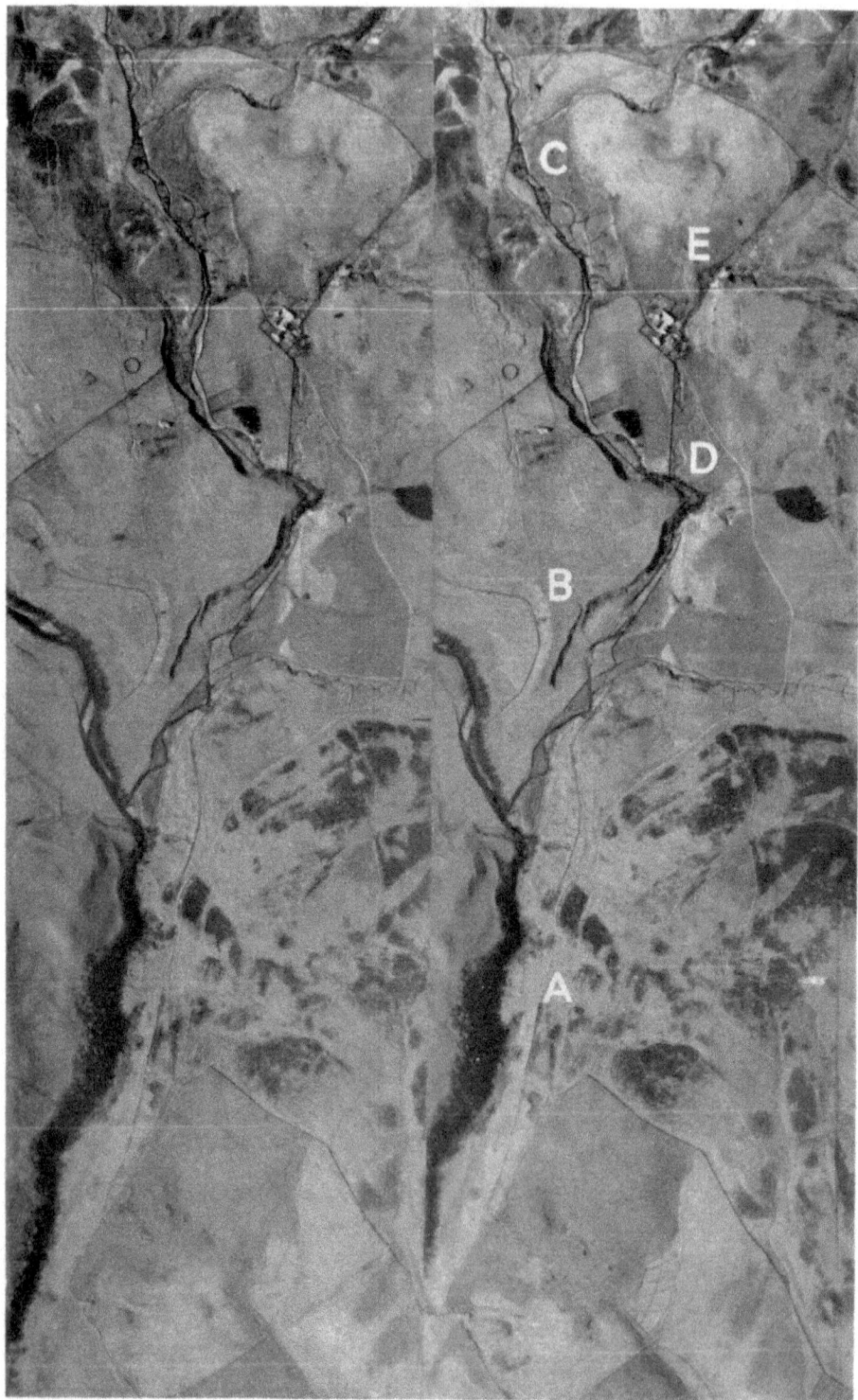

Plate 39

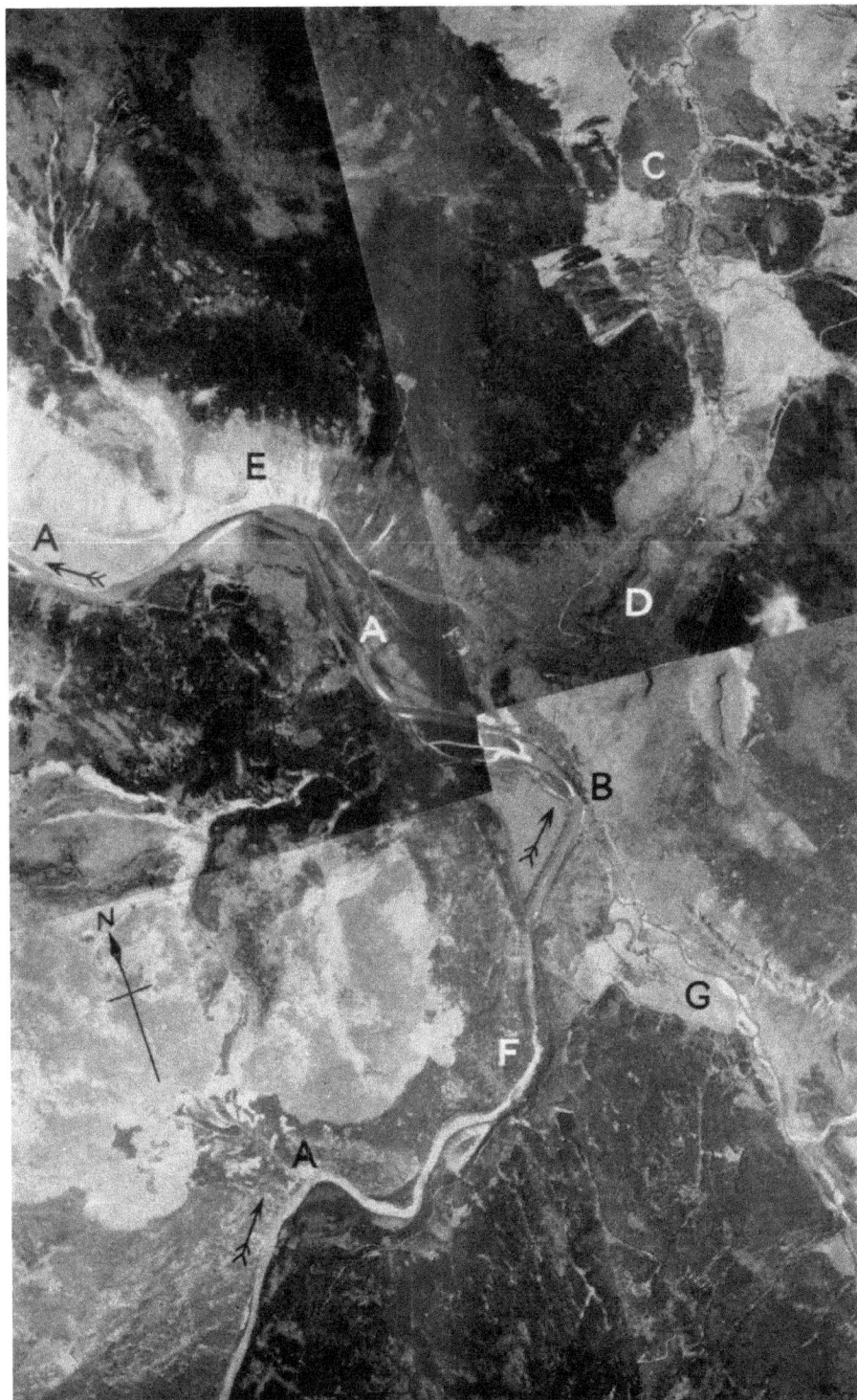

Plate 40

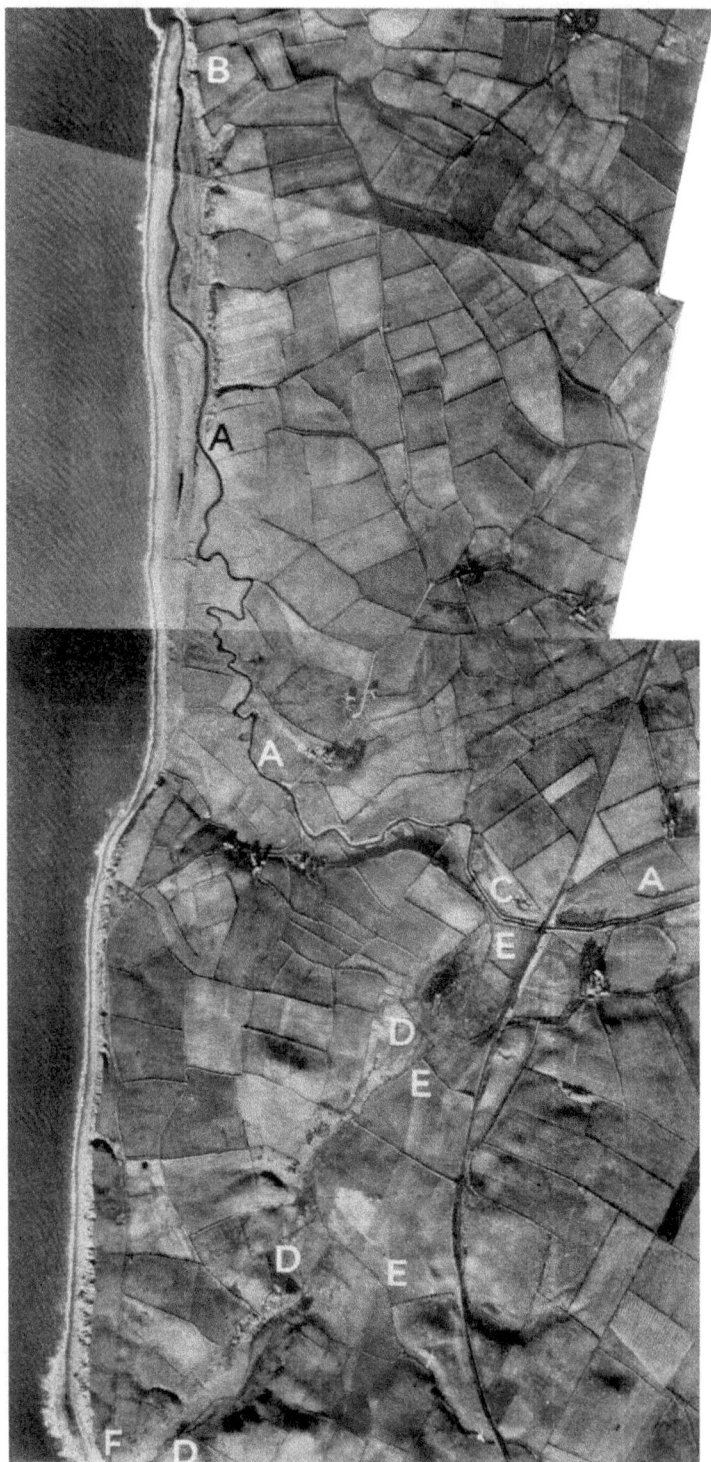

Plate 41

Plate 42

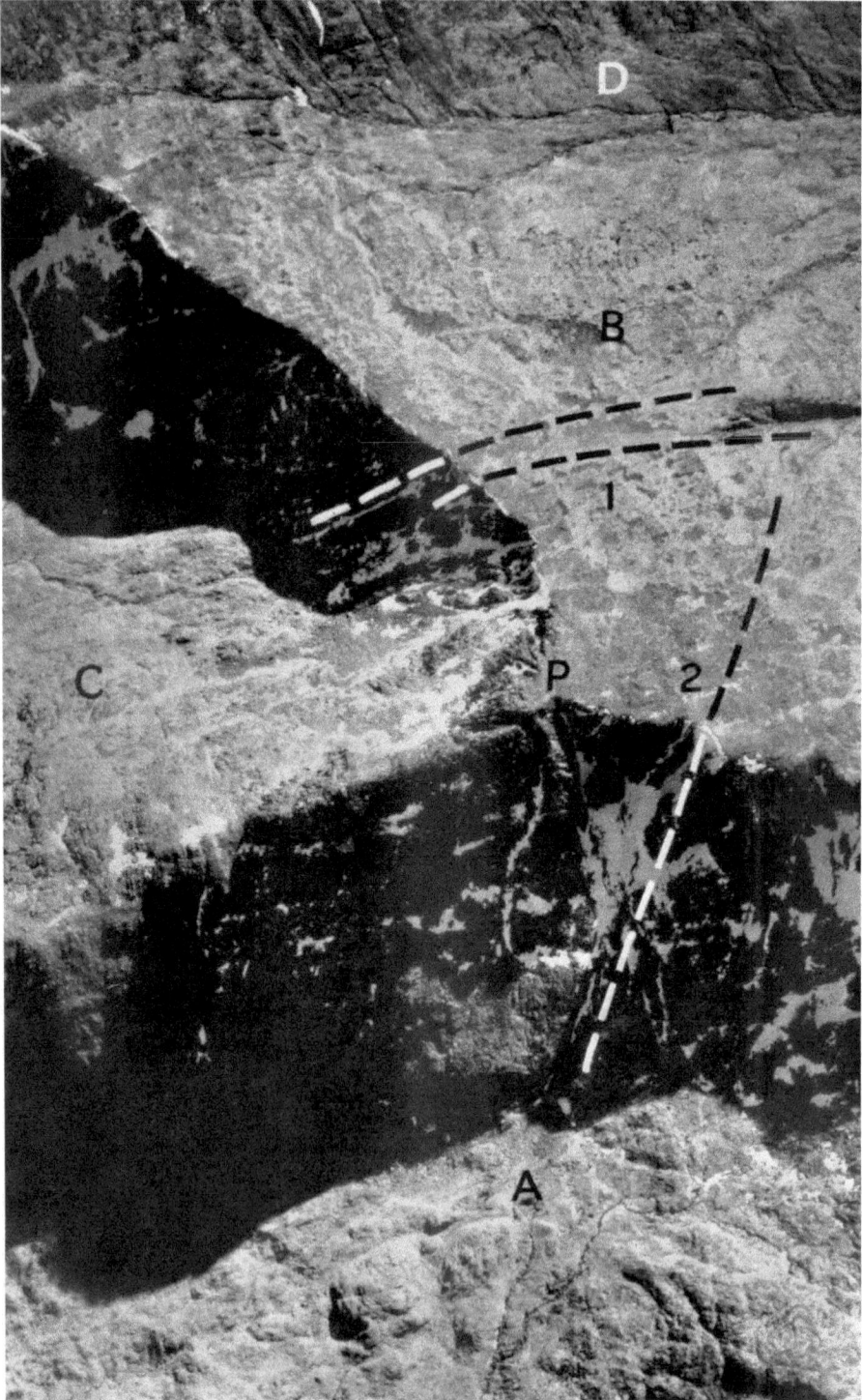

Plate 43

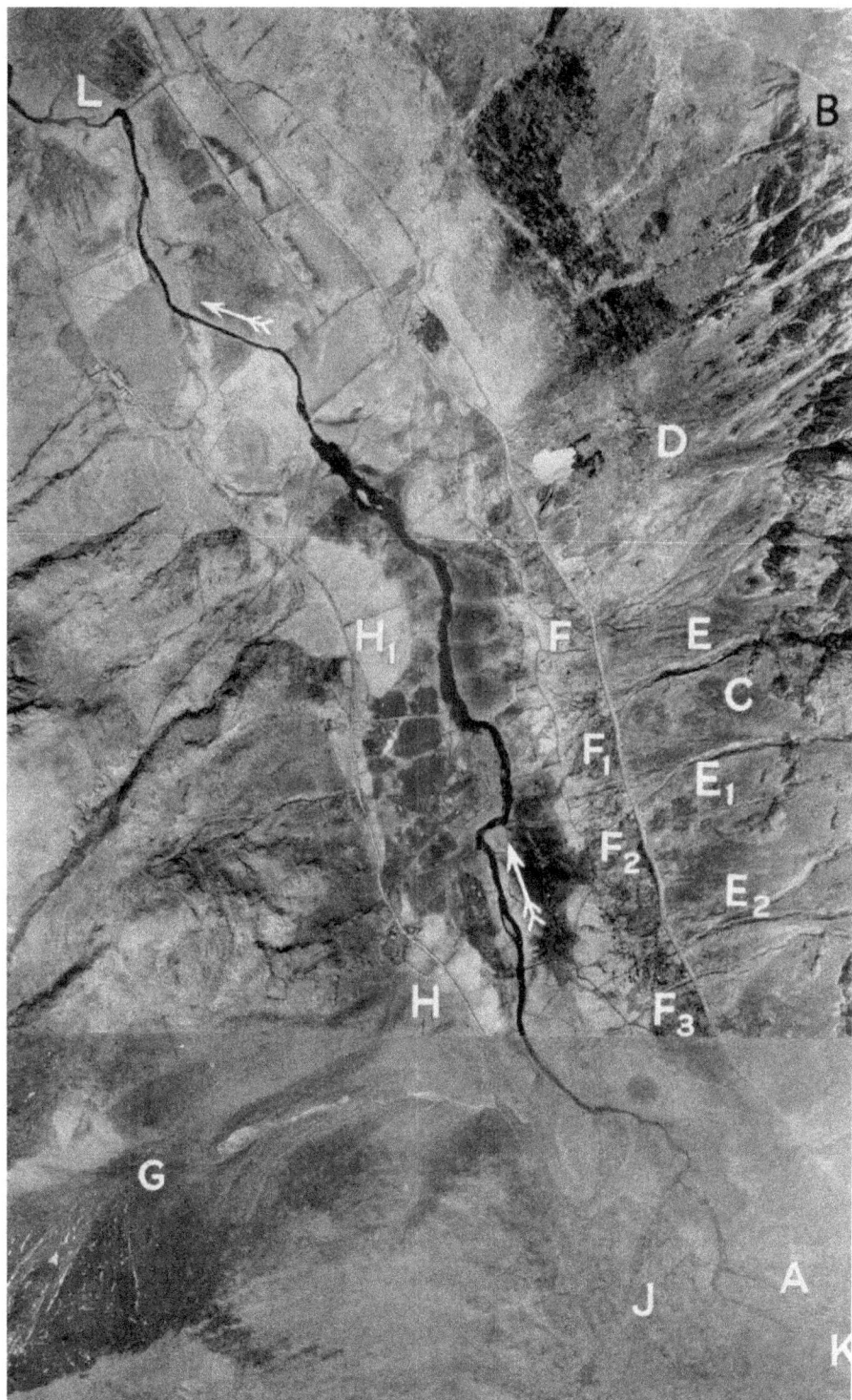

Plate 44

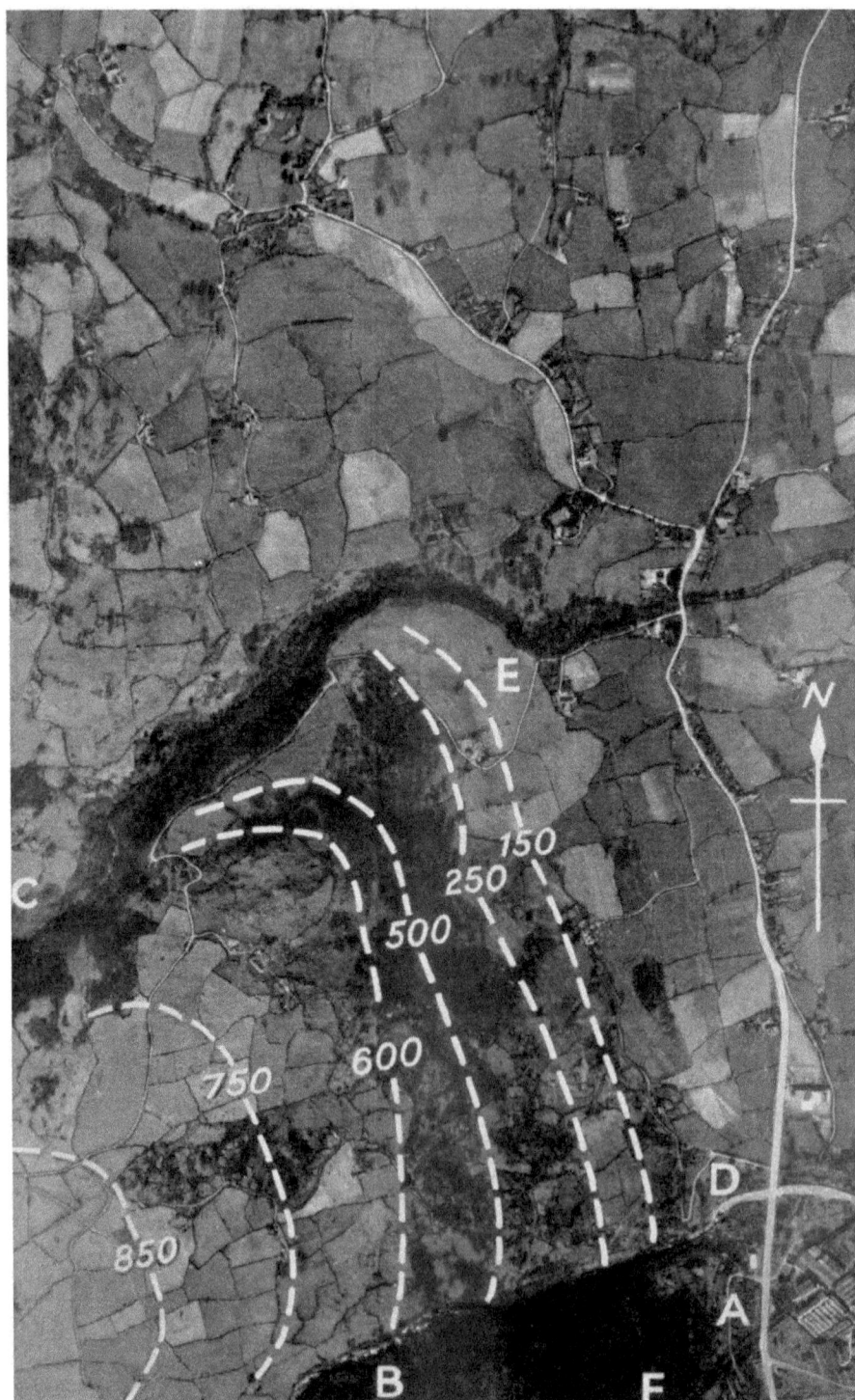

Plate 45

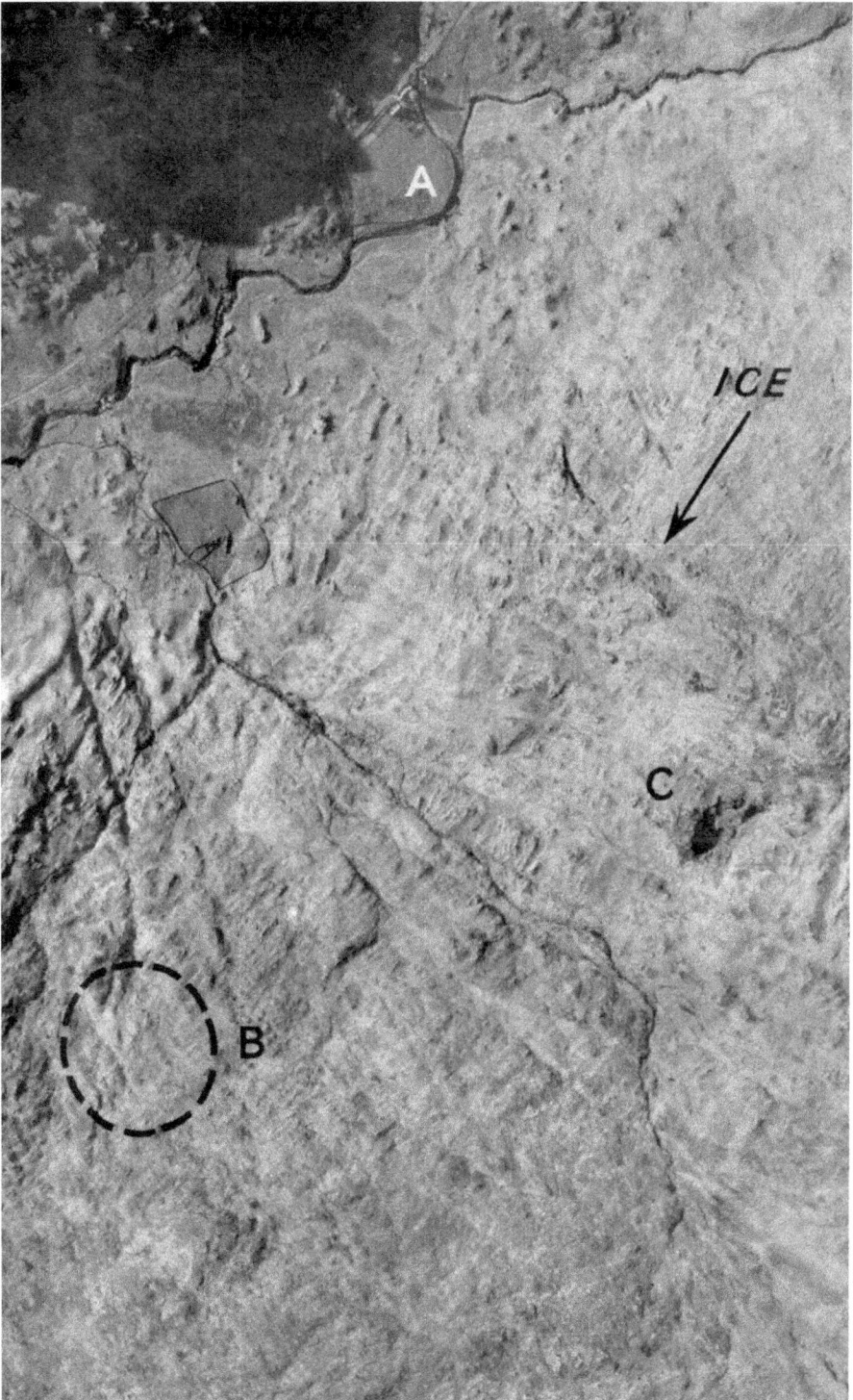

Plate 46

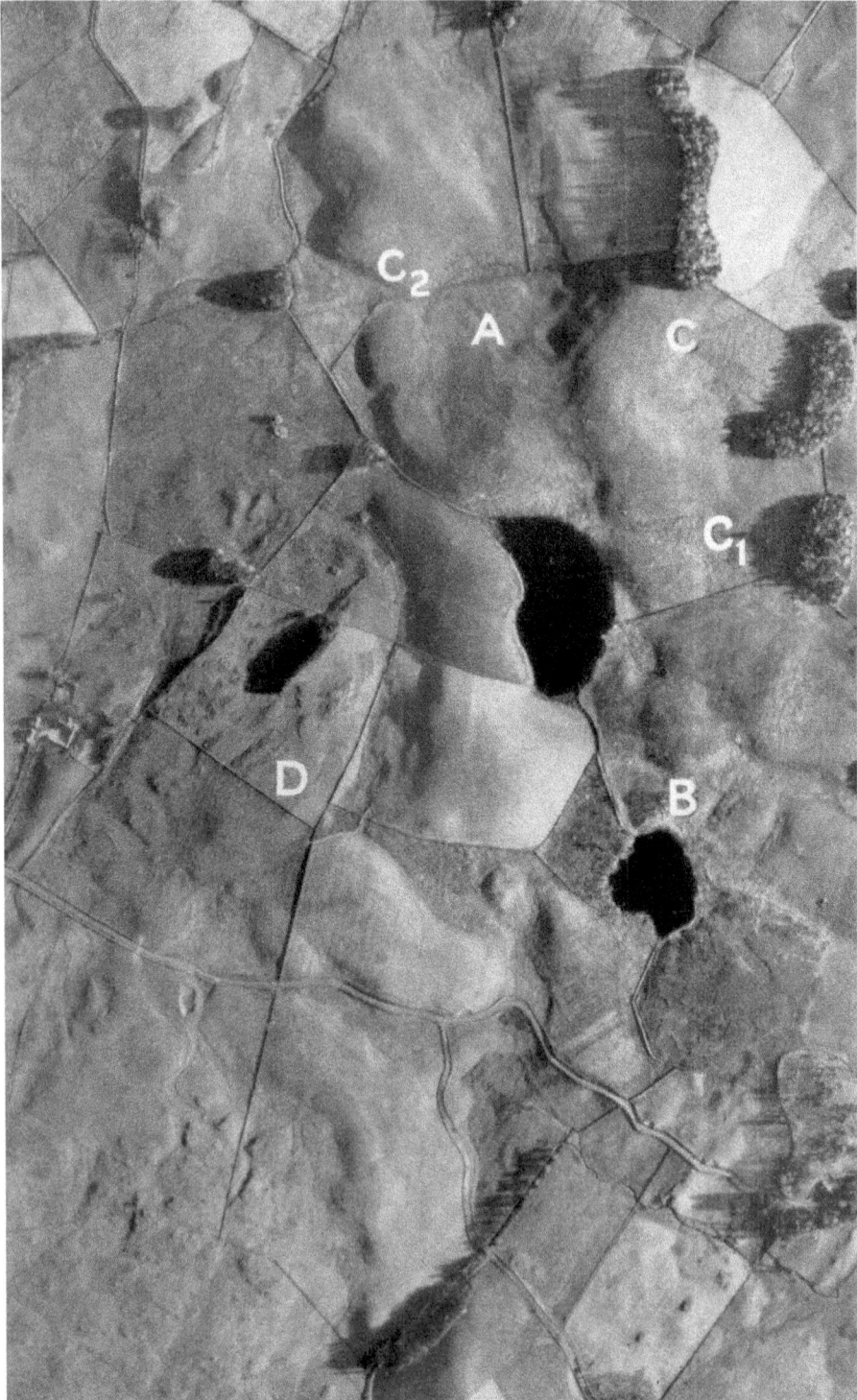

Plate 47

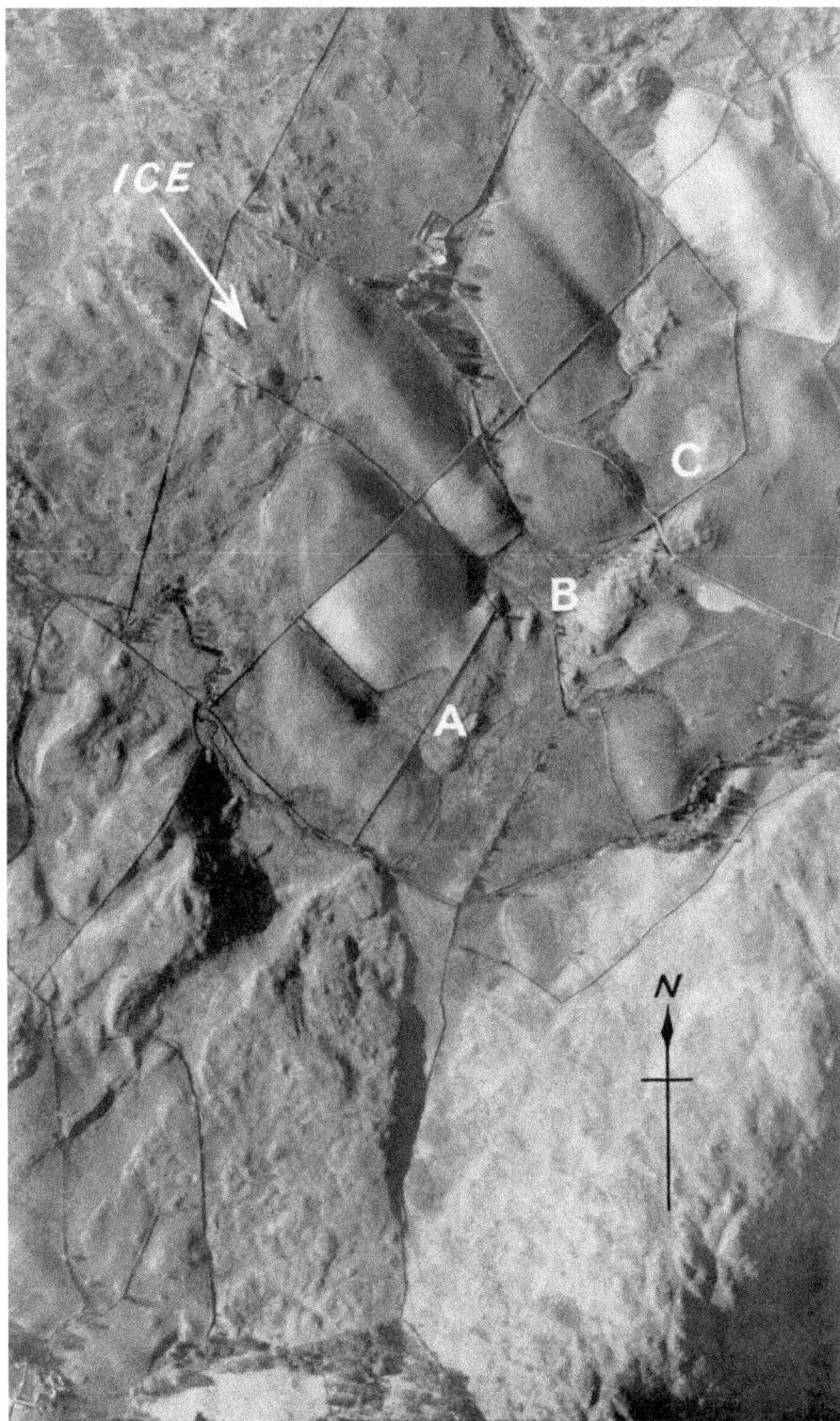

Plate 48

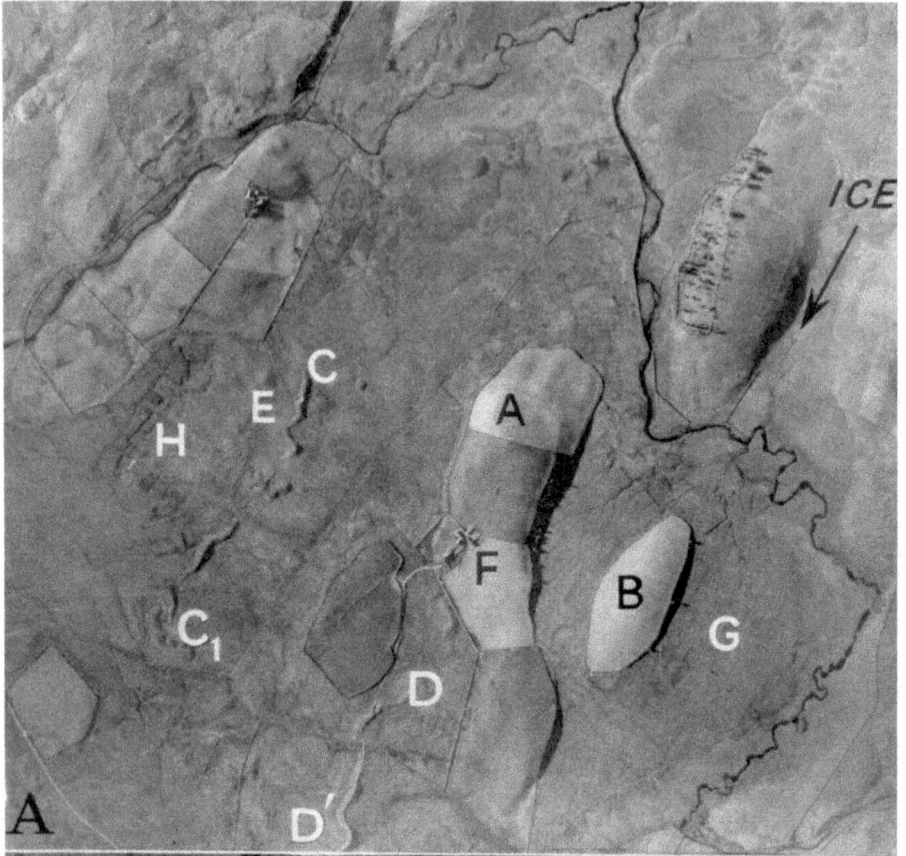

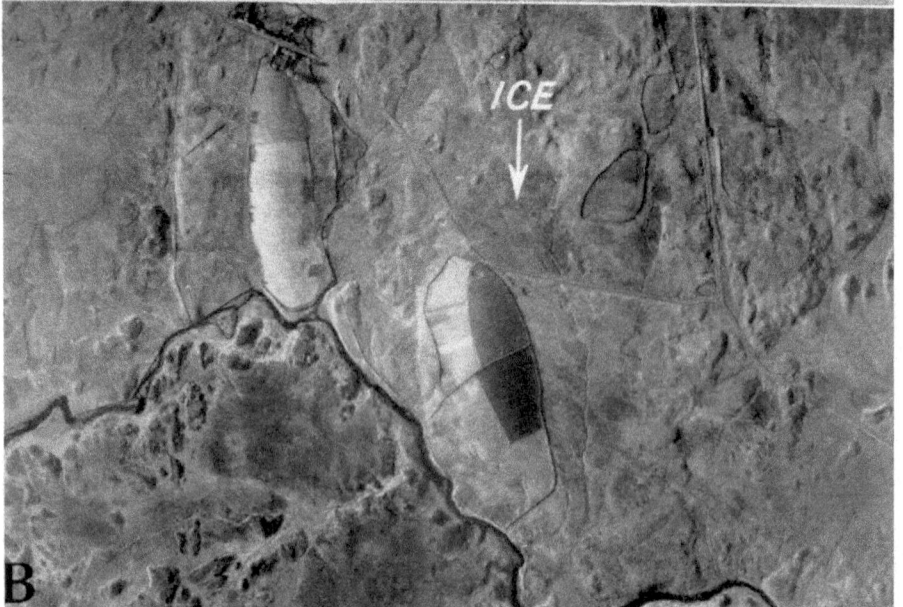

The corrie lake of Ffynnon Lloer itself is not now the regular or nearly round shape which often characterizes these rock basins or moraine-dammed lakes, but it seems quite probable from the photograph that its present irregular shape has been caused by the accumulation of debris from the higher slopes on its banks. Over the lip of the corrie at about 2,100 feet the lake is drained towards Llyn Ogwen and Afon Llugwy by the small Afon Lloer, but as yet this stream has not excavated an appreci-able notch in the lip of the corrie so that a prolonged period of erosion has to be envisaged before the lake will be drained and the corrie opened out like the long series of corries overhanging Nant Ffrancon on the West. The geology of the area shown on the photograph is moderately com-plicated since it includes Silurian material in the floor of the corrie itself and both contemporaneous and intrusive igneous rocks in the higher area, but the effects of these varied outcrops is significantly limited since it is the erosional features of the corrie and the gullies which dominate the geomorphology of this district.

Unlike the example of Plate 41, it is a reasonably common occurrence for a mountain area to have a number of corries developed on opposed faces. Under these circumstances the recession of the corrie walls will ultimately lead to the reduction of the highest portions of the mountain to knife-edge crests or arêtes and produce the jagged outlines which characterize many highly glaciated mountain regions. Idealized diagrams of this process normally show three such corries cutting back to form a 'sharp' mountain peak with three arête ridges radiating outwards and separating the three corries. Plate 42 is, in effect, a photographic repre-sentation of this state of affairs on Ben Cruachan, about twelve miles East of Oban, overlooking the Pass of Brander. (It should be explained that the photograph has been printed with its Northern edge at the bottom of the page so that the heavy shadows may fall towards the reader, an arrangement which enables many observers to appreciate relief forms on single air photographs more readily.) The three corries are A Coire Chat, B Coire à Bhachaill, and C Coire Dearg, and it will be appreciated that all three depart from the theoretical conception of a corrie as seen on Plate 41, because in each case the corrie has been widened out and lost its arm-chair shape, and the corrie lakes, if they once existed, have been drained and replaced by the headwaters of youthful streams, of which there is an excellent example at D. These changes are, of course, attribu-table to the fact that erosion has proceeded a great deal further in these cases than in the example chosen for Plate 41. Indeed, erosion has pro-ceeded so far that the three arêtes which combine at the summit of Cruachan (P., 3,689 feet) have a true knife-edge form, a feature which is beautifully picked out on the photograph by the fortunately isolated rims of snow lying immediately under the lee side of the knife-edges, and the mountain as a whole, seen from the Pass of Brander, has a highly distinctive peaked summit.

The overriding importance of this type of erosion in developing such a landscape is emphasized by the fact that the whole of this area is composed of the Cruachan Granite, a rock which in an unglaciated region or even a glaciated region without marked corrie development would normally tend to produce rounded uplands of the kind shown on Plates 9 and 45.

Within the main mass of the Granite a number of dykes cross the higher areas of Ben Cruachan, the group marked at (1) being basic dykes running in a general East–West direction, and that indicated by (2) a North–South dyke of Porphyrite. Though these dykes do produce minor surface features, it will be seen that the general pattern of erosion by corrie recession has not been materially affected by them.

In the area where a glacier has occupied its true valley tract the relief feature which is normally regarded as being most characteristic is that of an overdeepened valley with a U-shaped cross-section. Plate 43[1] was therefore selected as an example where the appearance of the valley on the Ordnance Survey map suggested a near approach to such a land form, since the contours on the valley sides are closely crowded and the valley floor is exceptionally level. It shows the upper and middle parts of the Nant Ffrancon valley which has undergone appreciably less post-glacial modification than most of the valleys in the surrounding areas of North Wales, and should therefore be reasonably typical of the theoretical U-shaped valley. In general, the morphology shown on the photograph does correspond with such a conception since the slopes above the roads to the East and West of the valley are precipitous, whereas the character of the river suggests that the plain itself is very nearly horizontal and in fact no contour line crosses the valley between points A and L. On the other hand the photograph does act as a corrective to the erroneous impression that the steep sides of glacial valleys are in fact the sheer faces suggested by the U-shaped sectional diagram. The appearance of the right-bank valley wall on the contour map indicates an exceptionally steep and straight slope, but it will be seen on the photograph that a considerable degree of modification has taken place. The upper slopes of

[1] The contrasting quality of the two prints from which Plate 43 is constructed arises from the fact that they are taken from the centre and starboard runs of the same sortie and the conditions of photography obviously differed between the two cameras.

this mountainous face have been gullied (see Plate 41) by a large number of watercourses, some of which can be seen at the crest of the slope on the present print at B. This stream action, combined with rapid weathering and possibly the presence of some morainic material, has led to the lower slopes being masked by immense accumulations of debris, including large angular rock fragments which can be seen at C and D, and through this scree-like slope the small streams are now excavating quite deep channels at E, E_1, E_2. Where these streams reach the flat valley bottom below the road the very abrupt change of gradient has led to the deposition of alluvial cones of rather indeterminate shape which appear as light patches on the photograph at F, F_1, F_2, F_3, and across which the streams spread out in a series of distributaries to flow into the darker coloured, ill-drained area between the cones and the main river.

The contrast between the right-bank and left-bank walls of the valley is brought out clearly on Plate 43. It will be seen on the map that the upper slopes of the latter were pitted with a long series of corries and one example of these can be seen in the South in Cwm Côch at G. Unlike the corrie of Ffynnon Lloer in Plate 41, these corries have all reached a fairly advanced stage of post-glacial erosion, they now contain no lakes, the corrie walls have been opened out and important streams flowing from them have eroded this side of the Nant Ffrancon valley into a much more irregular form than that on the opposite bank, and though it is still very steep it is obviously in the process of being broken down. Two of these left-bank tributaries have also built up alluvial cones where they join the plain, and in these cases the greater progress of erosion has produced at H and H_1, more highly developed cones than those at F, F_1, etc., the example at H_1 having a true cone shape, the formation of which has led to a highly characteristic diversion of the tributary round the edge of the cone to join the main river downstream.

The valley itself has the characteristics of straightness and the complete absence of spurs normally expected in glaciated areas, whilst its extreme flatness has already been mentioned. Of this the broad meandering river and the many traces on the photograph of ill-drained conditions provide ample evidence. It may also be noticed, however, that the tiny settlements avoid the valley floor and the two roads also keep to the foot of the valley walls and, in fact, mark the line of junction between the Silurian material of the valley sides and the alluvium of the floor, except in the South at F to F_3 and J where the valley sides close in and the roads climb steeply towards the upper parts of the valley head.

Here in the section of the valley above A there is an example of a further characteristic of glaciated valleys, that of the tendency for the long

profile of the valley to contain sharp irregularities referred to in some
cases as valley 'steps' and apparently caused by the fact that ice may, under
certain circumstances, erode unevenly and not simply towards a smooth
base line. Between *A* and *K*, in spite of the poor photographic quality, it
is just possible to see the white streaks which mark the presence of a series
of waterfalls where the river passes over one such step below the lake of
Llyn Ogwen. Thus, whereas the fall of the river across the whole of the
area shown on the photograph from *A* to *L* is exactly 50 feet, the fall in
the small stretch shown above *A* is between 150 and 200 feet.

Plate 44 shows the Western edge of the Vale of Conway to the North
of Dolgarrog (*A*) where it is suggested that a pre-glacial valley has been
modified by ice action and where conditions are much less diagrammatic
than in Nant Ffrancon. Nevertheless, the steep side of the valley remains
as an extremely important morphological feature and the form lines
inserted on the print to show the approximate position of the 150-, 250-,
500-, 600-, 750- and 850-foot contours demonstrate that in the area
between the Afon Porth-llwyd, flowing down to Dolgarrog, and Afon
Dulya to the North, the slopes below 600 feet are still abrupt. The
straightness of these form lines marking the end of the spur of higher land
between the two rivers may perhaps be taken to suggest that here there is
an example of the truncation of a spur by the development of the characteristically straight-sided form of an ice-modified valley.

The main interest of Plate 44, however, lies in the fact that it can be
used to study the post-glacial development of the two streams mentioned
above. Both streams occupied hanging valleys with an abrupt break of
slope or waterfall where this feature breached the steep wall of the main
valley. It is important to realize, however, that such hanging valleys in
this country seldom present the vertical lip, overhanging the U-shaped
valley, suggested by idealized block diagrams, since in most cases post-
glacial erosion has led to the notching of the valley wall by the tributary,
and to the upstream recession of the waterfall or break of slope. In both
the cases on the photograph it will be seen that in the zone corresponding
to the steep valley side, as indicated by the form lines, the streams have
cut narrow, gorge-like notches and that the waterfalls are now situated
beyond the rim of the main valley. On the Afon Porth-llwyd the series
of cascades and rapids can be seen quite clearly at *B*, but in the case of the
Dulya the falls near *C* are rather difficult to identify.

The availability of water power from the streams descending the steep
valley side was responsible for the existence of many mills along the
Western edge of the Vale of Conway, and two of the sites are marked at
D and *E*. A more modern version of this same phenomenon was created

by the construction of the aluminium works at Dolgarrog whose buildings just appear in the bottom right-hand corner of the print, though in
this case the head of water for the generators is derived by means of a
pipe-line (F) from a contour canal on the spur above the village rather
than by the direct use of the fall of the stream itself.

Where glaciers were not confined by traversing an actual valley but
moved over the intervening upland areas, the resulting land forms may be
extremely varied and depend in considerable measure on the local geological conditions and the effects of post-glacial erosion. One such case
has already been considered in Plate 12 where glacial and post-glacial
erosion accentuated the important relief forms arising from the nature of
the area's physical history. Plate 45, however, shows an area of uniform
rock character where the existence of an ice sheet has resulted in a hill
mass of smoothly rounded outline almost entirely devoid of soil cover,
though different interpretations may regard this as being due either to
ice scraping or to the absence of weathering beneath the ice. Here the
ice sheet moved in the direction shown by the arrow over the col
between Cairnsmore of Fleet and Millfore in Southern Scotland, parallel
with the valley of the Palnure Burn (A). In and near the valley there are
areas of glacial and fluvial deposition, but the North-western flank of
Cairnsmore which occupies the greater part of the print is composed
almost entirely of bare rock with the thinnest skimming of poor, light-
coloured soil round the ill-defined, post-glacial watercourses and isolated
groups of erratics, for example in the area outlined at B and around the
small rock basin at C. The whole of this area is composed of the Cairnsmore Granite and it is of interest to contrast the landscape here with that
of Ben Cruachan (Plate 42) where, on a generally similar rock, totally
different land forms have developed, largely because of the effects of
different types of ice action and post-glacial erosion.

The effects upon landscape of glacial deposition are seldom as striking
as those shown in the foregoing photographs, especially as the deposition
in many cases takes the form of large sheets of Boulder Clay and similar
materials which produce areas without well-defined relief features, where
vegetation and cultivation may hide the characteristics of the underlying
drift. Obviously no single photograph can demonstrate the very great
variety of surface and soil types associated with Boulder Clay, but where
the surface has the hummocky appearance with indeterminate drainage and
numerous lakes which are sometimes regarded as typical of little-modi-
fied Boulder Clay districts, these features are readily recognized on the air
photograph. Plate 46 shows one such area in South Scotland near the
village of Corsock about 9 miles North of Castle Douglas. Here there is

no defined pattern in the mantle of Boulder Clay, and the two lakes which lie in hollows in the hummocky surface of the deposit are characteristically not linked in any way with a systematic form of drainage. Most of the clay land has been under the plough and on some of the mounds it is possible to see the furrow marks, as for example near A, but the appearance of the lower area round lake B suggests the former ill-drained character of all the hollows in the drift surface, whilst the herring-bone system of drainage evident at C, C_1, C_2 indicates the steps which were taken to bring much of this area into cultivation. The appearance of the field D and the area round it arises from the emergence of small patches of solid rock through the mantle of drift.

Though the precise cause of the formation of drumlins and eskers is still subject to discussion, it appears to be reasonably certain that where considerable movement of sheet ice occurred the ground moraine may take on the much more varied character associated with the presence of these highly characteristic minor relief forms. Plate 47 shows an isolated area of glacial deposition surrounded by a region of mainly bare rock over which the ice sheet of the Ken glaciers moved South-eastwards parallel with the Ken valley mentioned above in connection with Plate 35. The area of deposition is composed almost entirely of a group of drumlins and it will be appreciated that these are aligned with the direction of ice movement. Apart from their characteristic shape and alignment the drumlins are distinguished from the surrounding country-side by the fact that they are under cultivation, possibly because of the better drainage of their curved surfaces, but it will be noticed again that it has been necessary to drain the low-lying parts of the drift artificially at A, B, C, etc.

Photograph A on Plate 48 shows an area near Drumabrennan, 7 miles West of Newton Stewart, which includes a group of drumlins, two of which are sufficiently large for their outline to be indicated by the contours of the 1:63,360 Ordnance Survey map of the area. Two of the drumlins (A and B) show up remarkably clearly because of their cultivation pattern and the fact that their limits have been defined precisely by the stone walls round the fields.

From C to C_1 and from D to D_1, the photograph also shows two esker ridges which, like the drumlins, have an alignment related to the direction of ice movement. It is interesting to notice that on the esker CC_1 a small excavation at E indicates the extraction of the sand and gravel of the ridge whilst the esker DD_1 has been used as the site for the track leading to Barnharrow Farm (F), presumably in order to keep the track above the level of the surrounding ill-drained clay land.

The extent to which drainage has been confused by the deposition of glacial material may be deduced from the extremely irregular course and varying width of the stream which winds through the lower areas between the drumlins. Throughout these areas the Boulder Clay clearly formed an ill-drained lowland—a fact which may be seen in the multitudes of small watercourses, the artificial drainage lines at G, and the general absence of cultivation away from the drumlins.

The rectangular pattern at H and the slightly different surface texture within it marks an area where cutting has taken place in the peat beds which overlie the drift throughout considerable parts of this region.

Photograph B on the same plate shows a small area near the hamlet of Knowe, 7 miles North-west of Newton Stewart, where two isolated drumlins are again rendered specially distinct by the way in which cultivation is confined to their rounded slopes. The main interest in the photograph, however, lies in the evidence it affords of the diversion of the River Bladnoch by the glacial deposits.

Where glacial deposition takes the form of corrie, valley or terminal moraines their recognition on air photographs is in large part dependent on the siting of the mound of material in relation to the relief forms with which it is associated. One example has already been included in the case of the corrie and upper valley moraines which appeared on Plate 31, and a further exceptionally clear illustration of the appearance of a morainic ridge is given by Plate 49. The photograph shows a short stretch of the valley of the Esk upstream from the village of Lealholm, and the area included within the dotted line marks the position of the terminal moraine of the Eskdale ice, whose height can be appreciated from the dimensions of the railway cutting (outlined at A) through it. Where the character of the moraine is not hidden by other topographical features or by the progress of cultivation the irregular surface to be expected in a mound of unsorted material deposited in situ can be easily recognized, especially in the area to the North of the railway line.

The glaciers which approached the Esk valley from the North and from its seaward end moved upstream as far as Lealholm where they appear to have reached the limit of their extension along the line of this moraine and to have formed a wall of ice blocking the whole of the valley at this point.[1] The existence of this ice wall of appreciably greater height than the moraine itself (the crest of which is at a height of 425 feet just South of the railway) meant that a considerable ice-dammed lake, referred to by Kendall as Lake Eskdale, was impounded in the portion of the Esk valley above the barrier. The waters of this lake appear to have escaped down the line of the present valley side along the Southern margins of the ice and

[1] See Kendall and Wroot, *Geology of Yorkshire* (1924), vol. 1, pp. 493-5, 503-4.

to have commenced the excavation of an overflow or meltwater channel at a height of 525 feet which can now be traced along the course of Wild Slack (*B-B'* on the photograph). A comparatively small Northward and Eastward withdrawal of the ice subsequently led to the exposure of a lower part of the moraine ridge over which the meltwaters began to pour and so excavated a second overflow channel at an initial height of 500 feet which was later eroded into the gorge-like stretch of Crunkly Gill from *C* to *C*$_1$. The opening of this second channel led to the abandonment of Wild Slack which is now a fairly typical dry meltwater channel, though its appearance on the photograph is confused by the vegetation. On the other hand the barrier of the moraine itself has so far prevented the development of a stream course along the lowest central portion of the valley, and the Esk continues to flow through the former overflow channel of Crunkly Gill which has therefore taken on more of the character of a normal river-worn gorge than a typical meltwater channel, though here again the thick woodland prevents a thorough study of its morphology.

Fluvio-glacial erosion may lead to an extremely wide range of minor relief features but one of the most interesting and distinctive land form associated with it is the meltwater or overflow channel which has been chosen as the subject for the last illustration in this section. The top part of Plate 50 shows a section of the Esk valley, a few miles downstream from the area shown on Plate 49, which was occupied by the ice of the Esk glacier. The wide spur of higher land marked by the form lines is the lowest part of Murk Mire Moor which lies between the Esk valley and the valley of the Murk Esk flowing Northwards just to the East of the area shown on this print. The meltwaters from Lake Eskdale and from the lakes impounded in the dales on the Southern side of the Esk have already been described as escaping along the Southern margin of the Esk glacier, and in this area they continued their flow by escaping across Murk Mire Moor and along the Western edge of the Murk Esk ice to join Newtondale and thence flow into the larger Lake Pickering in the South.[1] The overflow of these waters led to the excavation of channels across Murk Mire at successively lower levels as the ice retreated, and Plate 50 shows such a channel known as Moss Swang which developed during a postulated third stage of the ice withdrawal. The channel runs from *F* to *A* and has isolated a small portion of the moor outlined by the 600-foot form line. It will be seen that the channel has the customary U-shaped cross-section associated with the flow of extremely large volumes of water over a comparatively short period, whilst under the stereoscope it is possible to recognize the convex profile of the floor of the channel, its highest portion

[1] Kendall and Wroot, op. cit., pp. 494–501, including map.

being between D and E. In addition to the main outlet at A it may be noticed that there is a subsidiary channel at B at a higher level to the West of Castle Hill (C) which may represent the course of the overflow during an intermediate stage before the ice retreated sufficiently to allow of the use of the outlet at A. These channels, now high on the side of the moor, are of course almost entirely devoid of streams, and this absence of normal water erosion and the fortunate absence of woodland means that their morphology can be studied most clearly by the method of air photography.

Chapter 4

MINOR RELIEF FEATURES AND SOILS

It was indicated in Chapter 1 that one of the advantages of the use of air photographs for geographical purposes lay in the fact that they may be used to demonstrate and study minor relief features and variations of soil character which frequently have great significance in the human geography of a region. A considerable number of such minor features have already been discussed in so far as they can be related simply to geological or erosional considerations, but in addition Plates 51 to 56 are included to illustrate a number of land forms and distributions which were not properly susceptible of classification in one of the foregoing main sections.

Plate 51 covers the edge of the Fenland area of Eastern England at the village of Thurlby (A), North-west of Peterborough, where a very slight slope up to the 50-foot contour (indicated on the photograph) is associated with the change from the superficial deposits of the fens to the Oxford Clay to the West. The main road (BB) coincides exactly with the junction of the Clay and the first of the fen deposits, the Fen Gravel, which extends from the road to the dotted line CCC. Beyond CCC the much darker tone of the photograph indicates the transition from the lighter coloured Gravels to the black soil of the reclaimed Peat land stretching beyond the limit of the present prints. The extreme flatness and low-lying character of the Peat areas may be appreciated from the spot height of 13 feet, some 20 miles inland, and the presence of raised embankments on either side of the Glen River crossing the corner of the photograph at D.

The Gravels appearing in this area have been described[1] as beach gravels marking the inland limit of inundation of the Fenland region, and it will be seen on Plate 51 that the Car Dyke EEEE, the canal constructed round the perimeter of the Fens during Roman attempts at reclamation, was excavated along the line of these Gravels, slightly above the true fen land, presumably in order to divert inland water into the main river channels and so prevent its draining into the fens themselves.

Apart from the difference in soil colour there is little on the photograph to suggest important differences in agriculture between the Peat and Gravel areas, though the latter shows the regular rectangular field pattern of reclaimed land to a much less marked degree than the former and does support some tree growth, unlike the greater part of the fens. To the

[1] S. D. J. Skertchley, *Geology of the Fenland* (1877), pp. 183-6, map p. 208.

West of the road, however, the contrast is much more marked, not only in the greatly increased proportion of grass-farming on the Clay, the presence of woodlands and hedges and the small irregular field shapes, but also in the much greater density of settlement outside the limits of the former ill-drained area. Round the Western edge of the fens, particularly to the North of Peterborough, there is a continuous belt of settlement in the form of villages at exceptionally regularly spaced intervals of little more than a mile along the slightly higher ground which was followed by the Roman road forming the Northward extension of King Street. Most of these villages are markedly elongated in form either along the line of the North–South road itself or along roads crossing it at right-angles and running from the higher ground towards the Car Dyke and the fens. Thurlby (A) and Northorpe (F) are reasonably good examples of this settlement form and distribution, though reference to the Ordnance Survey map of the area shows settlements of even more elongated form a little farther to the South at Langtoft and Boston. The very juxtaposition of two such villages is particularly interesting, however, in view of their names. Northorpe obviously means Northern 'thorpe', a termination normally taken to indicate a 'smaller village due to colonization from a larger one',[1] usually an adjacent 'by' village such as Thurlby, both terminations being derived in this case from Old Danish forms common in Eastern England.

Plate 52 shows an area of somewhat similar relief forms in the Somerset Levels. The top right-hand corner of the print shows a portion of the reclaimed levels of the basin of the River Ken, the partly canalized stream along the edge of the photograph being the Little River. The bottom left-hand portion of the photograph is also formerly ill-drained alluvium in the basin of the Yeo River, and here too the field drains and one major artificial drainage channel at A can be recognized. Between these two parts of the levels a low ridge of land marks the position of a tongue of Trias on which is situated Yatton (B) and North End (C), the distribution of settlement as well as the lighter tone of the photograph making the presence of this very small relief feature amply evident. (The gentleness of the morphology of the 'ridge' is realized if one considers the very small cutting at Yatton station which takes the completely level stretch of railway track through this feature.)

This zone of slightly higher, dry land across the ill-drained levels stretches from the high open area of the Carboniferous Limestone in the South at Ball Hill to within less than 2 miles of the Carboniferous Limestone of the Clevedon ridge, both of which are the site of a number of

[1] E. Ekwall, *English Place Names* (Oxford, 1936), p. 447.

prehistoric earthworks, including two large hill forts. It appears possible, therefore, that the Yatton ridge may have acted as a link between two ancient areas of settlement, and certainly since very early historic times the road and villages along it have been of some considerable importance. If the name Yatton is not the product of mis-spelling it is in all probability derived from the O.E. 'geat-tun' or 'tun in the pass', a description which fits its topographical setting remarkably well if 'pass' is used in one of its normal early senses of a 'passage' possibly between areas of flooded land.

This area of Somerset also provides a number of excellent examples of the phenomenon of outliers which formed extremely important 'islands' of dry land within the frequently flooded marshlands and mosses of the levels. Some of these, such as Glastonbury, for example, were the site of ancient settlements which have developed into modern towns, but Plate 53 has been chosen to show the Northern side of Brent Knoll where the absence of such urban development makes it rather easier to appreciate the distribution of settlement in terms of the relief feature itself. The nearly circular road from the village of Brent Knoll (A) to East Brent (B) delimits the outlier almost exactly. Outside this circle the Lower Lias Clays underly the superficial deposits of the levels, whereas within it the plateau-like area of the Knoll is composed of the Middle Lias Marlstone, and above this again the Upper Lias forms an almost conical hill (unfortunately only partially included in this photograph at C) on the crest of which a prehistoric hill fort occupies an extremely fine strategic site. After prehistoric times settlement concentrated on the lowest slopes of the Marlstone where communications were easy, water supply abundant (note the position of the waterworks at D and the ancient mill site at A), and varied agriculture possible using the summer meadows of the levels on the one hand and the relatively fertile soils of the Marlstone on the other. The intensive use of the Western and Southern slopes of the Marlstone platform, particularly for orchards and gardens, almost certainly reflects the importance of aspect and slope as is evident also on the near-by Westward and Southward facing slopes in the Mendips (see Plate 17).

A morphological type not dissimilar to the outlier is the residual hill, and one example of such a feature is represented on Plate 54. It shows one of the areas in which a hill mass composed of the Middle and Upper Chalk rises some 350 feet above the general level of the plateau of the Lower Chalk in the Marlborough Downs, about 10 miles West of the town of Marlborough at the White Horse (E), and it clearly marks an area where the presumably once-continuous cover of the higher elements of the Chalk has not, as yet, been completely eroded. The dotted lines mark the outcrop of the Upper Chalk (A), Middle Chalk (B) and Lower Chalk (C), and it will be seen that the outcrop lines have the appearance of form lines, especially round the Northern end of the hill where the direction of the shadows makes appreciation of the surface morphology particularly easy. The limits of the hill mass are most sharply defined on the North because of the great distinction between the arable land use on the Lower Chalk and the uncultivated poor grassland on the Middle and Upper Chalk, though it should be noticed that this abrupt change does not take place on the Lower Chalk in the South where the surface is broken by the typical deep coombe-like valleys leading down to Ranscombe Bottom. The area D marks a capping of Clay-with-Flints which overlies the Upper Chalk on a number of these residual areas, and it is of interest to notice the slightly different vegetation associated with this formation even when it occupies so small an area.

An additional reason for the choice of this particular example lay in the fact that it demonstrates the selection of the crest of this isolated hill mass as the site for the construction of the prehistoric camp of Oldbury Castle. It is of interest from an archaeological as well as a morphological point of view that whereas there is a double ditch and rampart round the greater part of the camp's circumference, only one such earthwork was regarded as necessary along the side of the camp which runs along the crest of the steepest slopes of the hill on the North-west. The long shadow in the North-west corner of the earthwork is cast by the slender monument which is a distinctive landmark throughout the Marlborough Downs.

One of the smallest forms of surface morphology is that produced by the tendency for the surface soil to slide, or rather creep, down land with steep gradients and in doing so to form a series of wrinkles or furrows, usually at right-angles to the slope. This soil creep, unless accentuated by former cultivation, tends to produce features which often have an amplitude of only a few inches, though they may be appreciably larger in well-developed examples. In the case of the very small furrows, recognition of this feature on air photographs is really only easy where the shadows are long and fall more or less along the line of the soil movement. The two examples on Plate 55 have been chosen as typical of a case where recognition is made easy by convenient shadows and one where little help is given by the shadows. Photograph A shows a portion of Rattledown Hill and East Dundry (South of Bristol) where a minor hill area on the Inferior Oolite overlooks the East Dundry valley on the Lower Lias Clay. The soil creep at C is taking place on the side of a small valley in the face of the Oolite and in consequence the shape of the 'folds' has been modified by the valley form so that an excellent impression is given by the photograph of the surface 'puckering' that is produced by this process, especially as the long shadows are falling across the line of the plications. On the other hand, photograph B shows a case where the shadows are less dense and the furrows much smaller. It shows two rounded valleys in the Yorkshire Wolds (adjacent to the area shown on Plate 16) and careful examination will reveal in each case a banded effect on the valley sides which is fairly easily recognizable at D and E. Here the direction of the folds is normal since they run nearly parallel with the contours of the valley sides and are not being influenced by any factors other than the downhill gradient.

The last land form to be included in the present section is the delta—in this case a lake delta. Though such forms perhaps belong more properly to the section on coastlines, the coastal examples of deltas are not readily demonstrated except by very large photographic mosaics unsuitable for the present volume. The photograph shows part of the Stroan Loch near New Galloway which has acted as the base level of erosion for a stream which is now nearly completely graded in its lower course and is building out a small delta into the lake, one side of which has a very characteristic finger-like extension. Two other unrelated points of interest on the photographs are first the excellent impression it gives of moorland country with isolated rock outcrops and a very patchy vegetation of heather and bracken, and secondly the somewhat unusual phenomenon of an isolated drumlin at A, partially enveloped in river alluvium.

Chapter 5

COASTS AND SHORELINES

The study of coastlines by means of air photographs suffers from the fact that most of the features which distinguish particular types of coast tend to be too large to be shown on individual photographs at large scales. For purely descriptive purposes this difficulty can be overcome by the use of small-scale photography (say 1:50,000), but when detailed study is necessary stereoscopic examination of the large-scale photographs, followed by the construction of mosaics for photographic reduction to a more manageable scale, is probably the best solution. In the present section Plates 57 to 64 have all been prepared from photographs of an approximate scale of 1:10,000, which has been retained in the case of Plates 58, 59, 63, 64, reduced to 1:15,000 in the case of Plates 57, 60, 62, and to 1:25,000 in Plate 61, which is included partly as an example of an unrectified mosaic which has undergone considerable reduction of scale before reproduction.

To offset this disadvantage the vertical air photograph has the great virtue of making many stages and processes in coast and shore development more easily understood since it 'fixes' such transitory features as the direction of wave attack, the consequences of wave refraction, and the appearance of purely temporary beach phenomena. Moreover, where repeated photography can be arranged it is possible to record pictorially the particularly rapid changes which take place in coastal areas, especially where deposition and the building of spits and bars is taking place.

The effects of structural trend lines on coastal forms are naturally immensely varied, and it is only possible to include two examples of the simplest cases where the trend lines result in longitudinal and transverse coasts respectively.

Plate 57 shows two adjacent portions of the coast of the Tayvallich peninsula overlooking the Sound of Jura and a little way South of Loch Crinan. Photograph *A* shows the island named Eilean Dubh and the adjacent mainland, and photograph *B* includes Carsaig Island and Carsaig Bay. On the coast and immediately inland the trend lines are those of the Caledonian orogeny which run from North-east to South-west in this locality and produce very varied relief features parallel with the coast. Here in the Tayvallich peninsula the Schists or Phyllites of the Dalradian metamorphics are sometimes replaced locally by igneous material, usually

63

Plate	Locality	Ordnance Survey Sheet no. 1:63,360	Latitude and Longitude	Sortie number	Print numbers
49	Lealholme near Whitby	86	W 00°50' N 54°27'30"	106G UK 1700	4189
50	Near Goathland North Yorkshire	86	W 00°45'30" N 54°25'30"	106G UK 1700	1226
51	Thurlby near Stamford	123	W 00°21' N 52°44'30"	CPE UK 1932	4244 4246
52	Yatton Station Somerset	165	W 02°50' N 51°23'30"	CPE UK 1869	4142
53	Brent Knoll Somerset	165	W 02°57' N 51°15'30"	CPE UK 1869	4315
54	Oldbury Castle near Calne	157	W 01°55'30" N 51°25'30"	106G UK 1415	4069
55A	East Dundry near Bristol	165	W 02°36' N 51°23'30"	CPE UK 1869	3133
55B	Uncleby near York	98	W 00°45' N 54°01'30"	106G UK 1313	3135
56	Stroan Loch New Galloway	Scotland 87	W 04°08' N 55°00'30"	106G Scot UK 52	3198
57A	Eilean Dubh Sound of Jura	Scotland 60	W 05°39'30" N 56°01'	58/A/433	5039
57B	Carsaig Island Sound of Jura	Scotland 60	W 05°38'30" N 56°02'	58/A/433	5188
58	Aros Bay Islay	Scotland 69	W 06°02' N 55°41'30"	LEU UK 7	7250
59	Coast East of Dunnet Head	Scotland 12	W 03°19' N 58°39'	106G Scot UK 133	3087 3088 3089
60A	Scarborough	93	W 00°24' N 54°17'	106G UK 394	4004
60B	Coast East of Sidmouth	176	W 03°11' N 50°41'	106G UK 1412	3217
61	Rattray Head	Scotland 31	W 01°51' N 57°37'	106G Scot UK 107	3001 3003 4001 4003 4005 4007
62A	Pagham Harbour	181	W 00°45' N 50°45'30"	541/217	3031
62B	Dawlish Warren	176	W 03°26' N 50°36'	CPE UK 1824	3074 3075 3076
63	Coast North of Ballantrae	Scotland 82	W 04°59' N 55°08'	106G Scot UK 150	4122
64	Gillespie Burn Luce Bay	Scotland 90	W 04°43' N 54°50'	106G Scot UK 42	4186

Plate 49

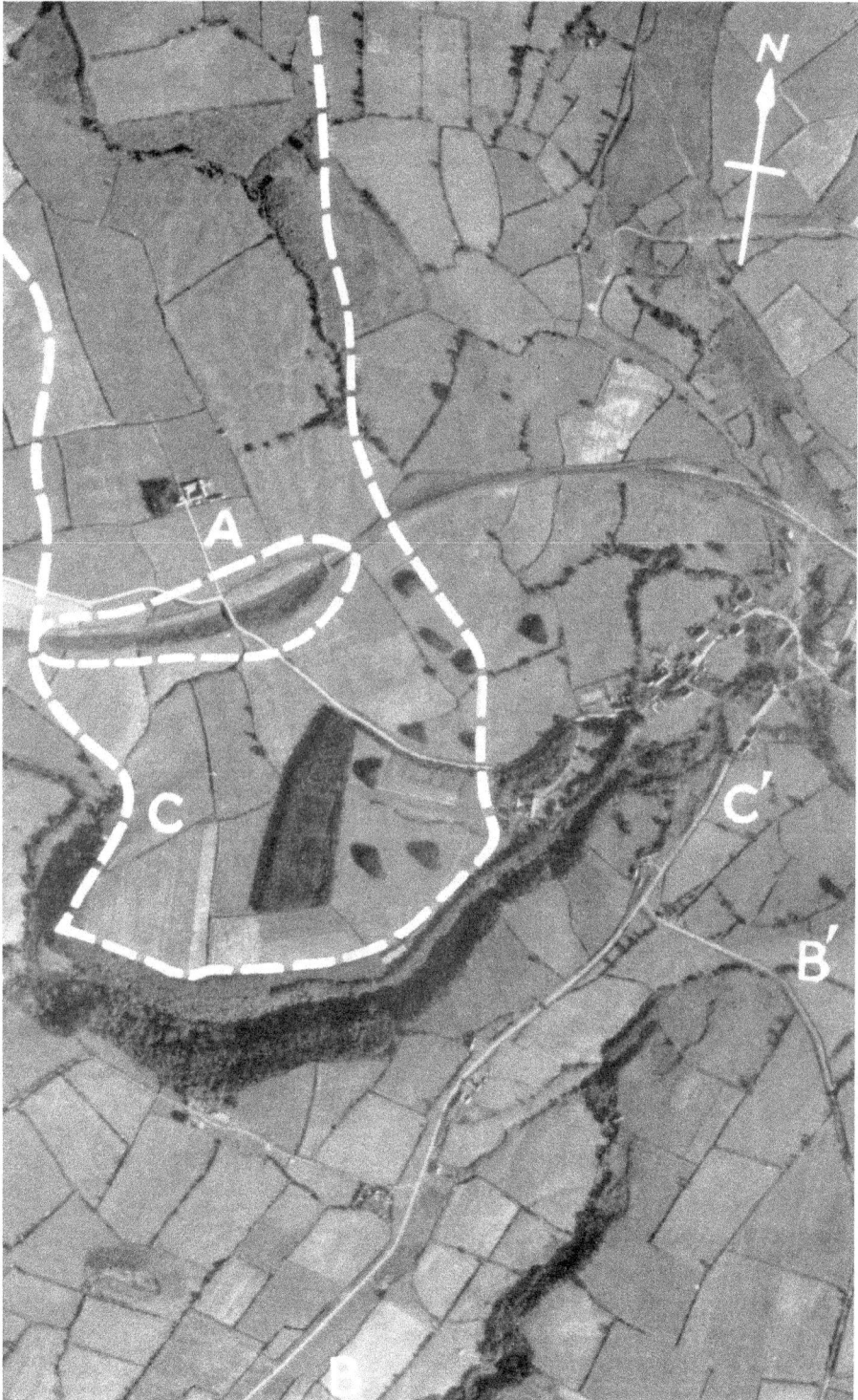

Plate 50

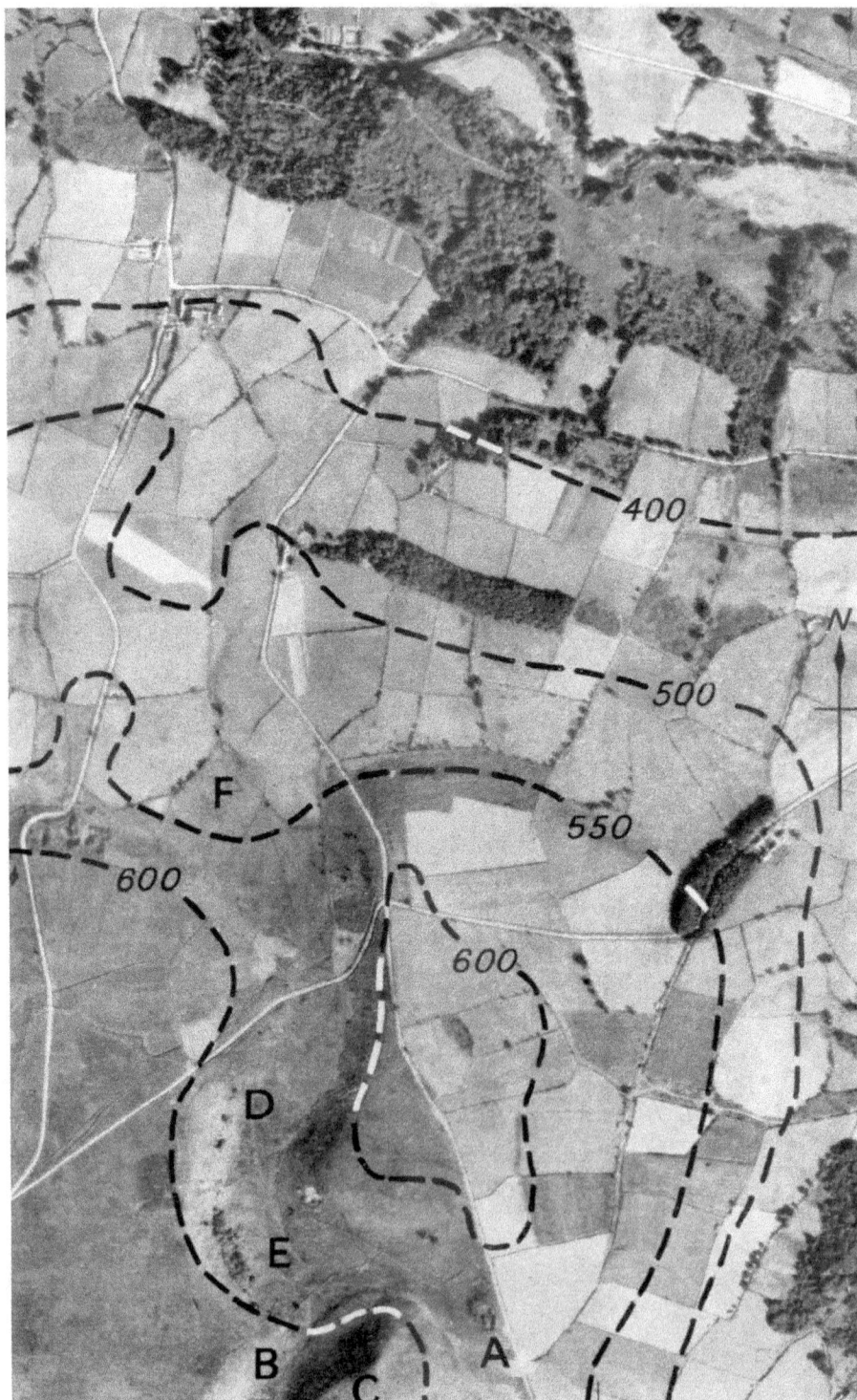

Plate 51

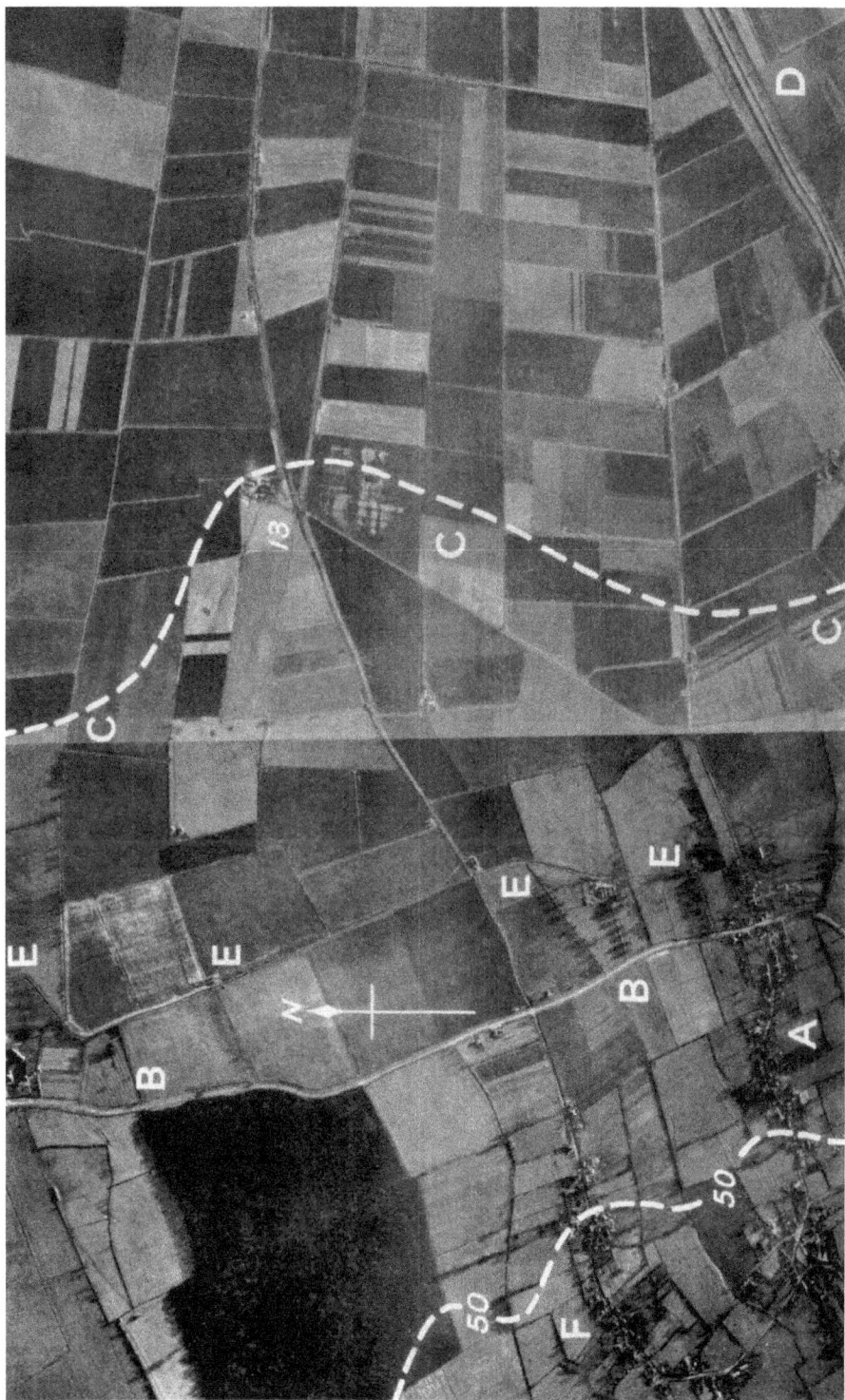

Plate 52

Plate 53

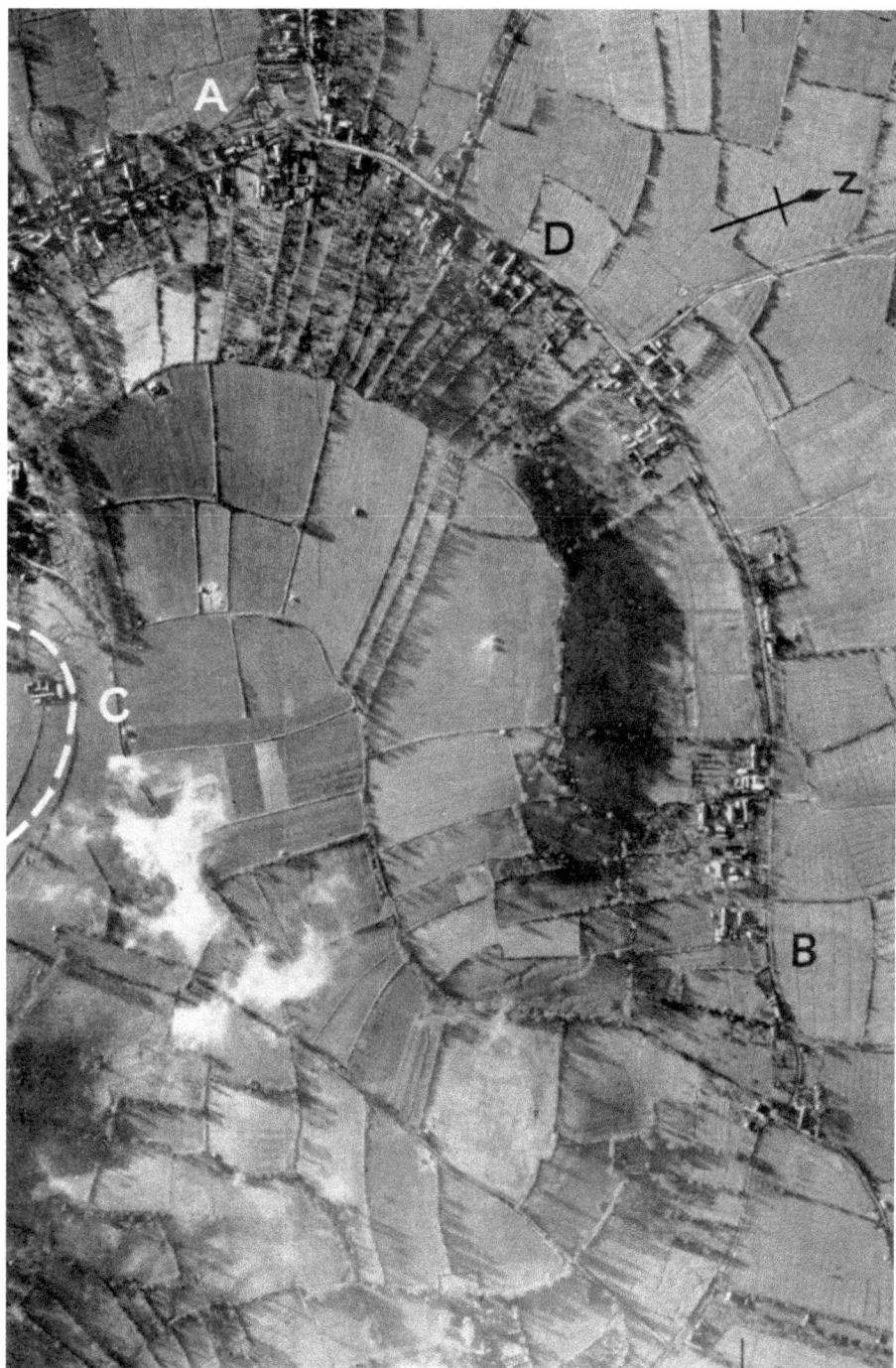

Plate 54

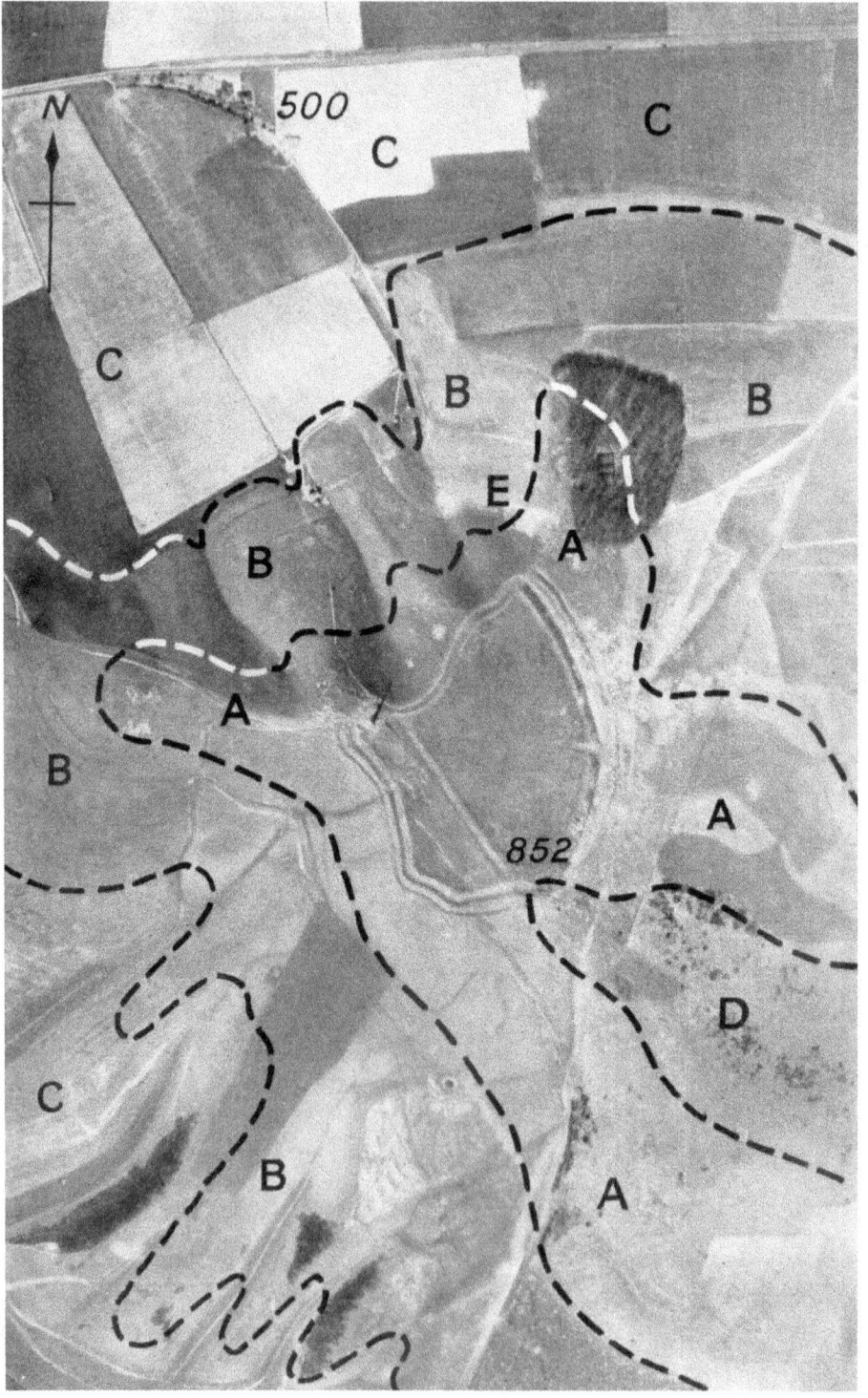

Plate 55

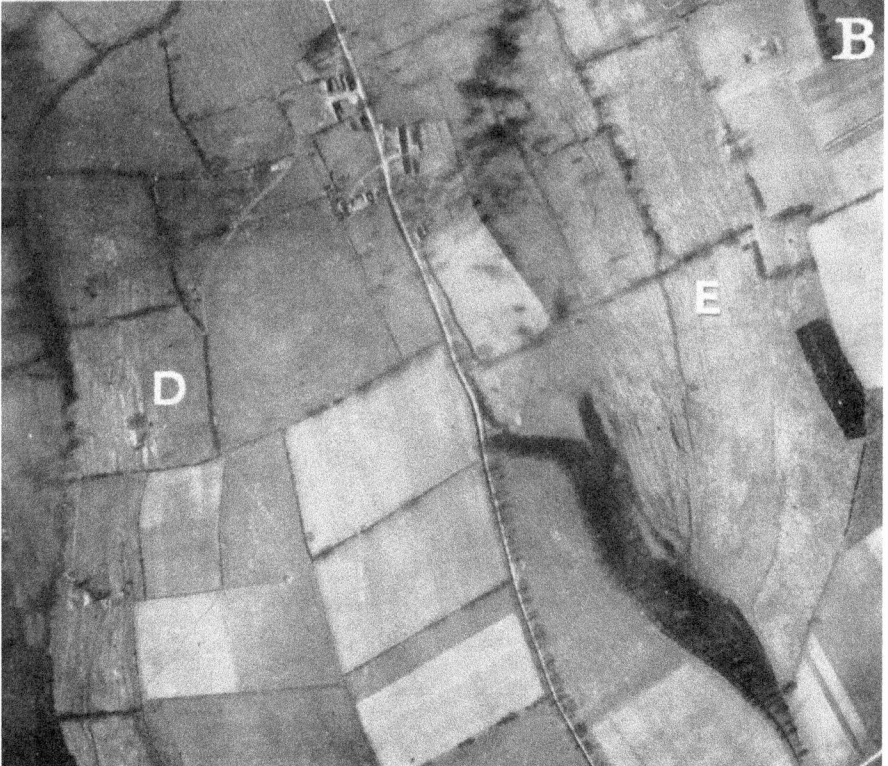

Plate 56

A

Plate 57

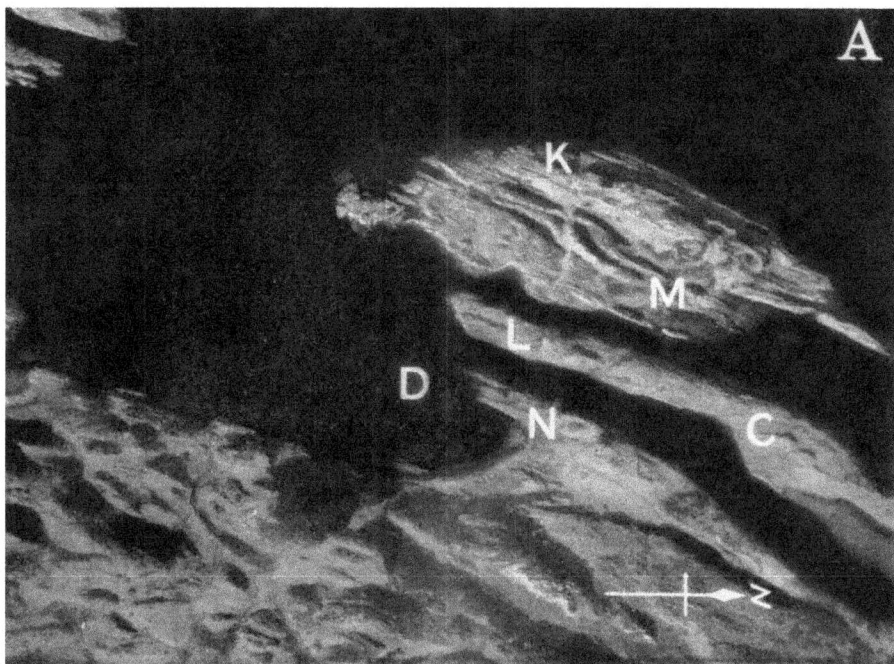

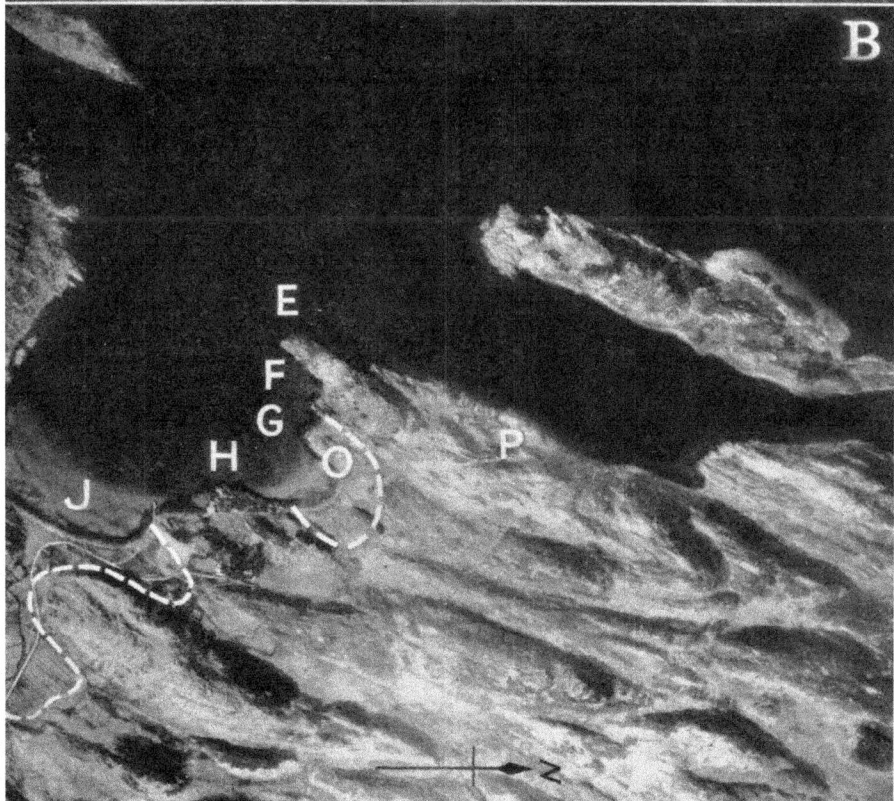

Plate 58

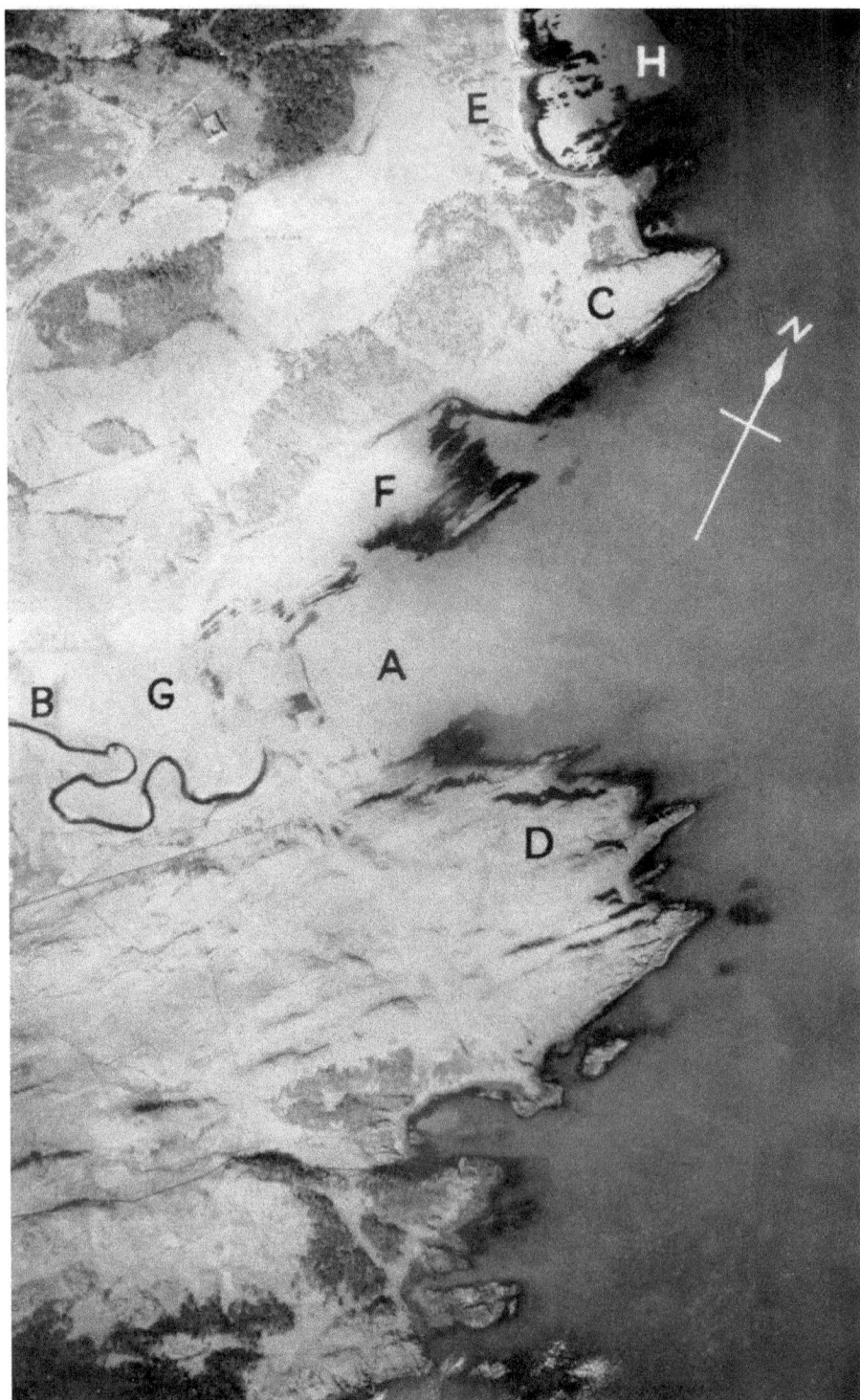

Plate 59

Plate 60

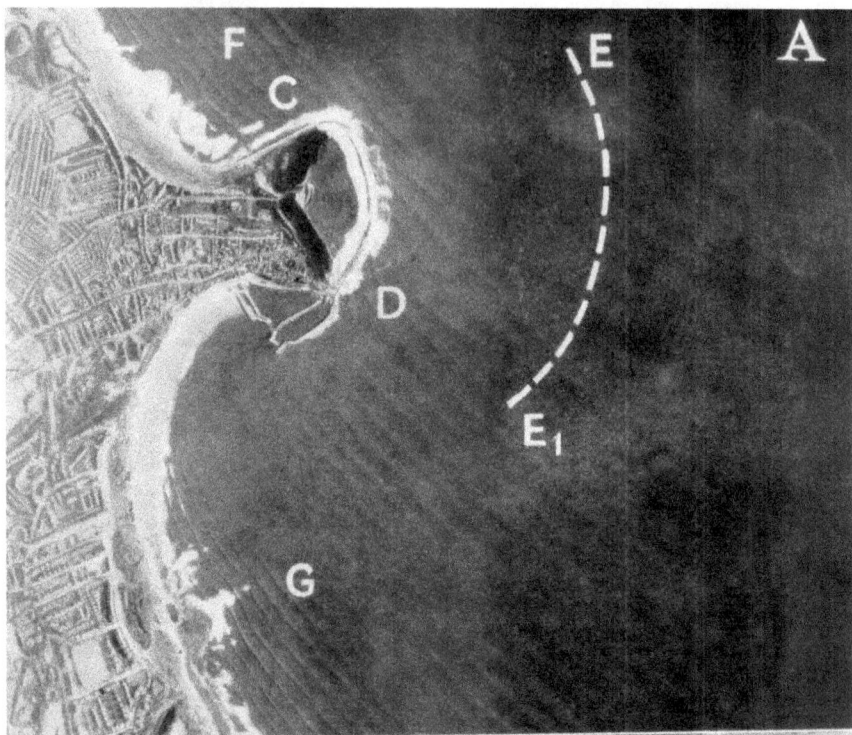

Plate 61

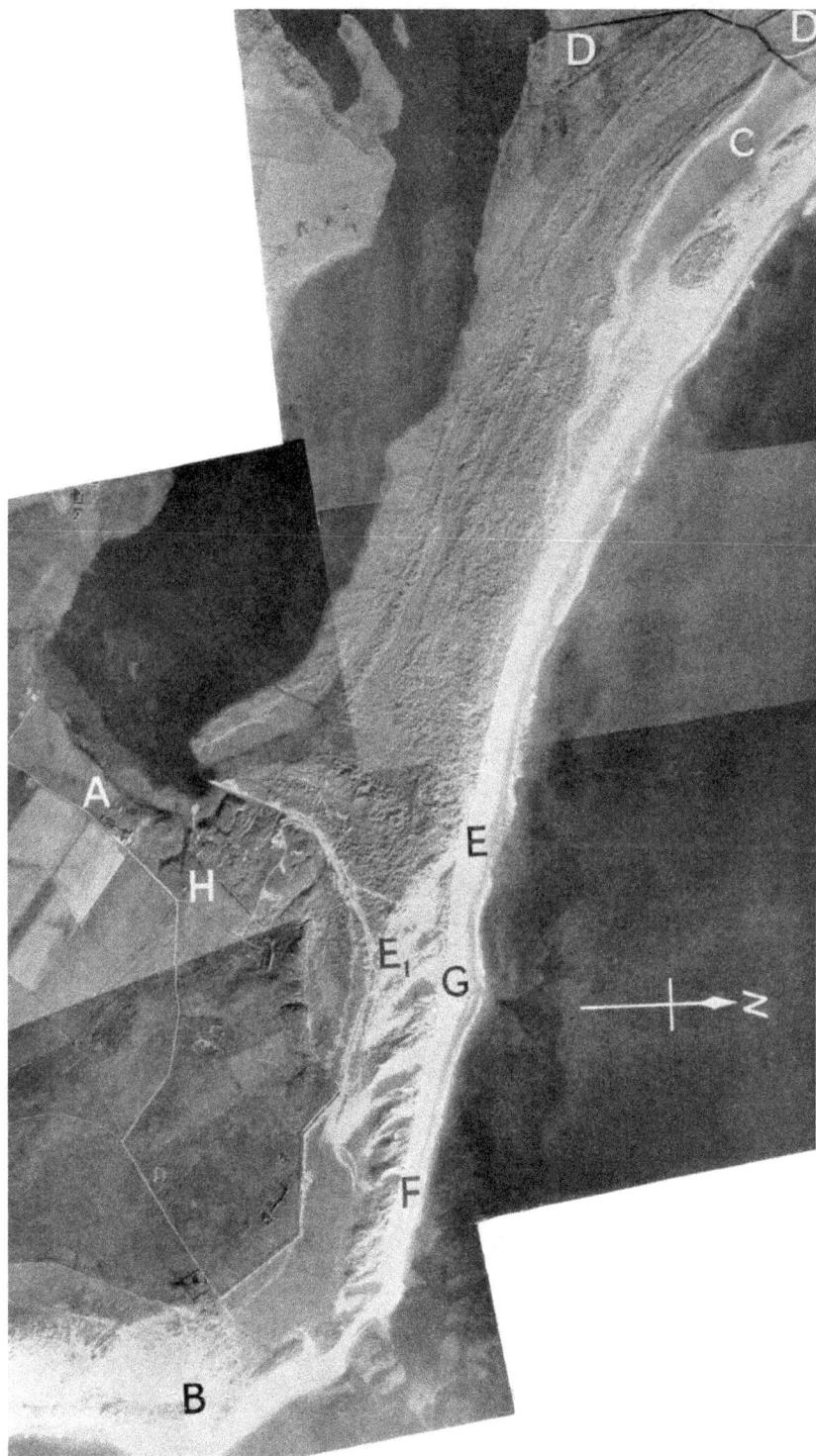

Plate 62

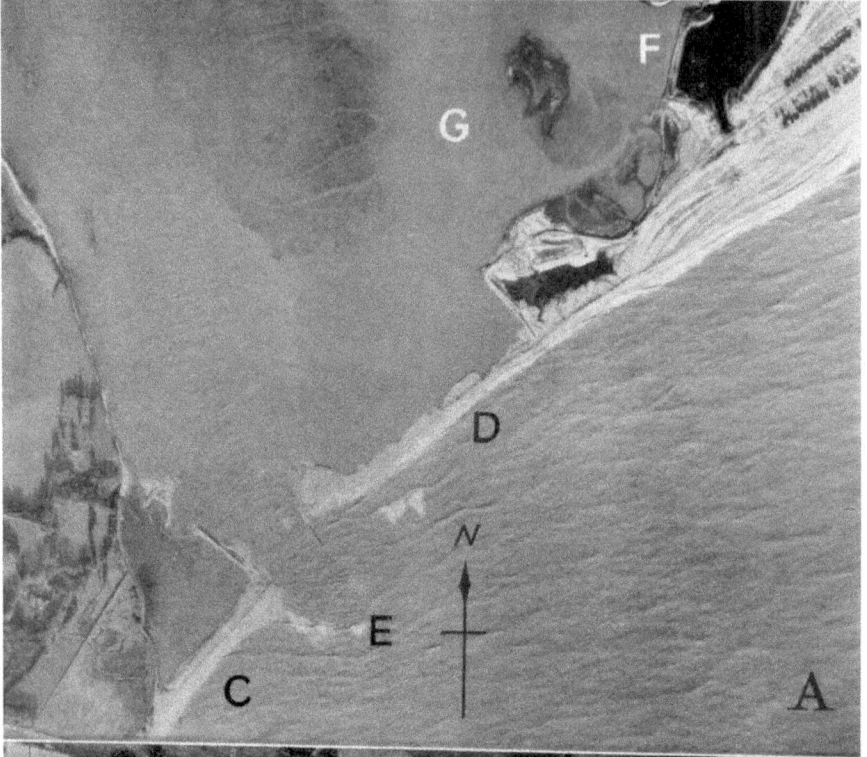

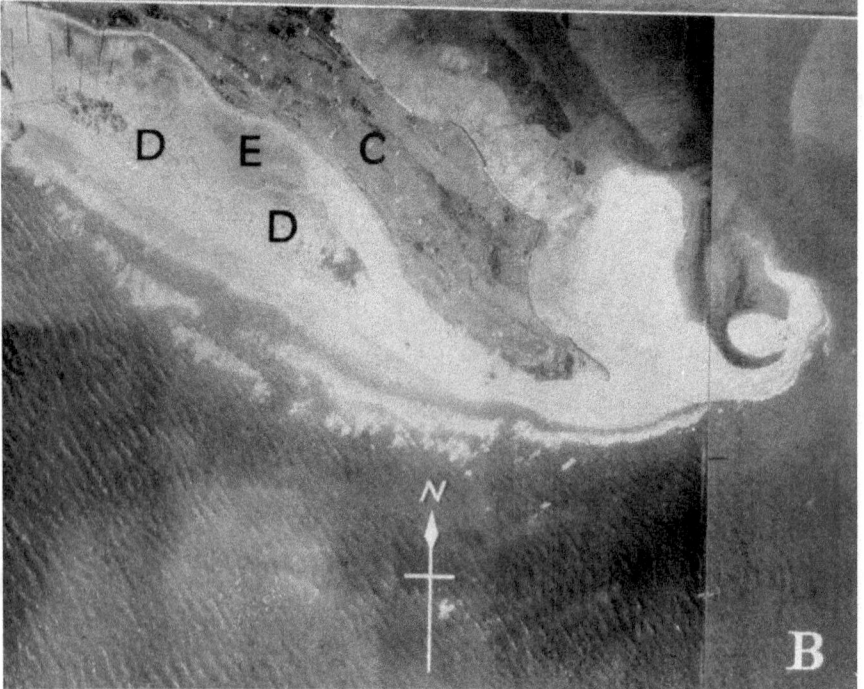

Plate 63

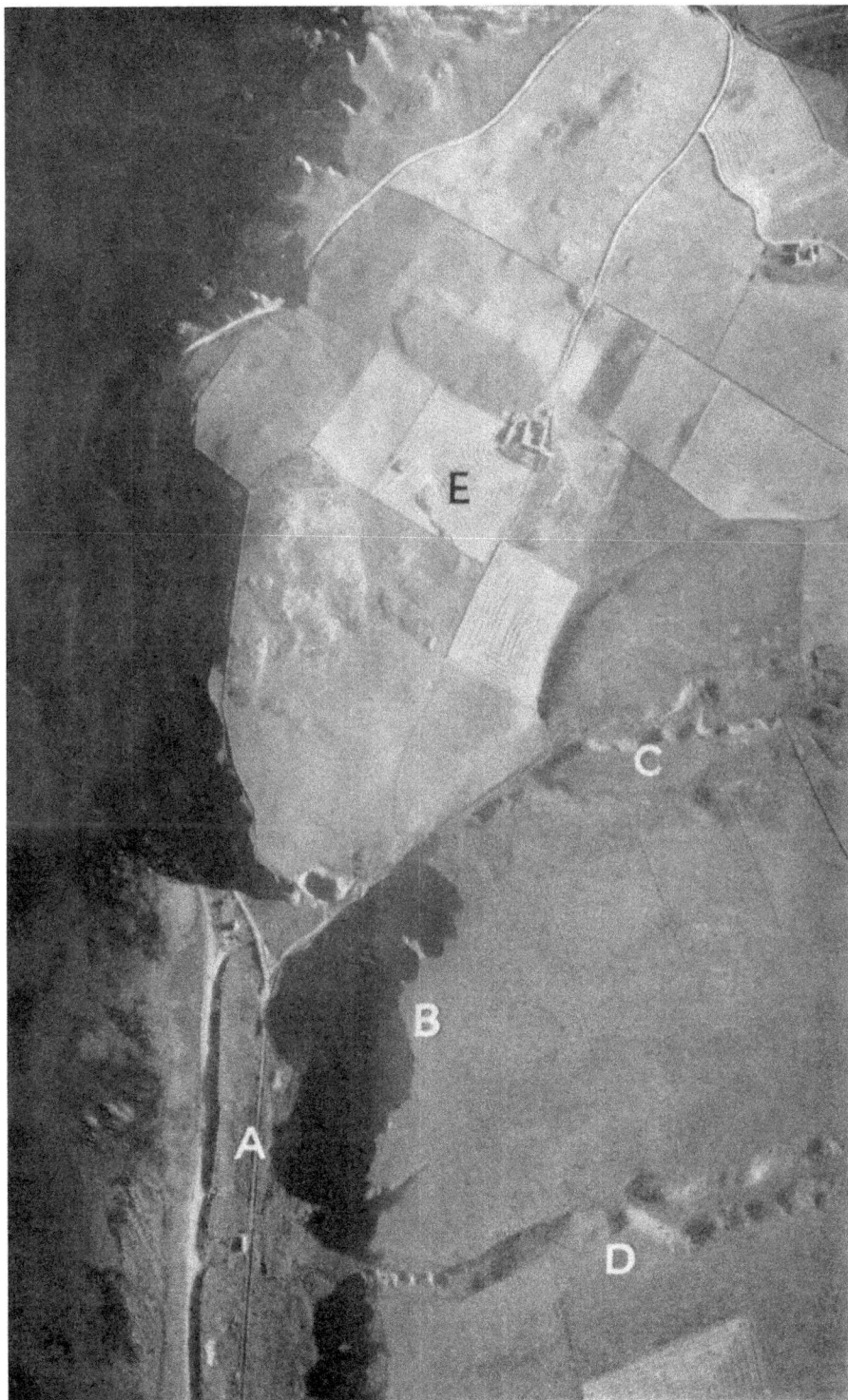

Plate 64

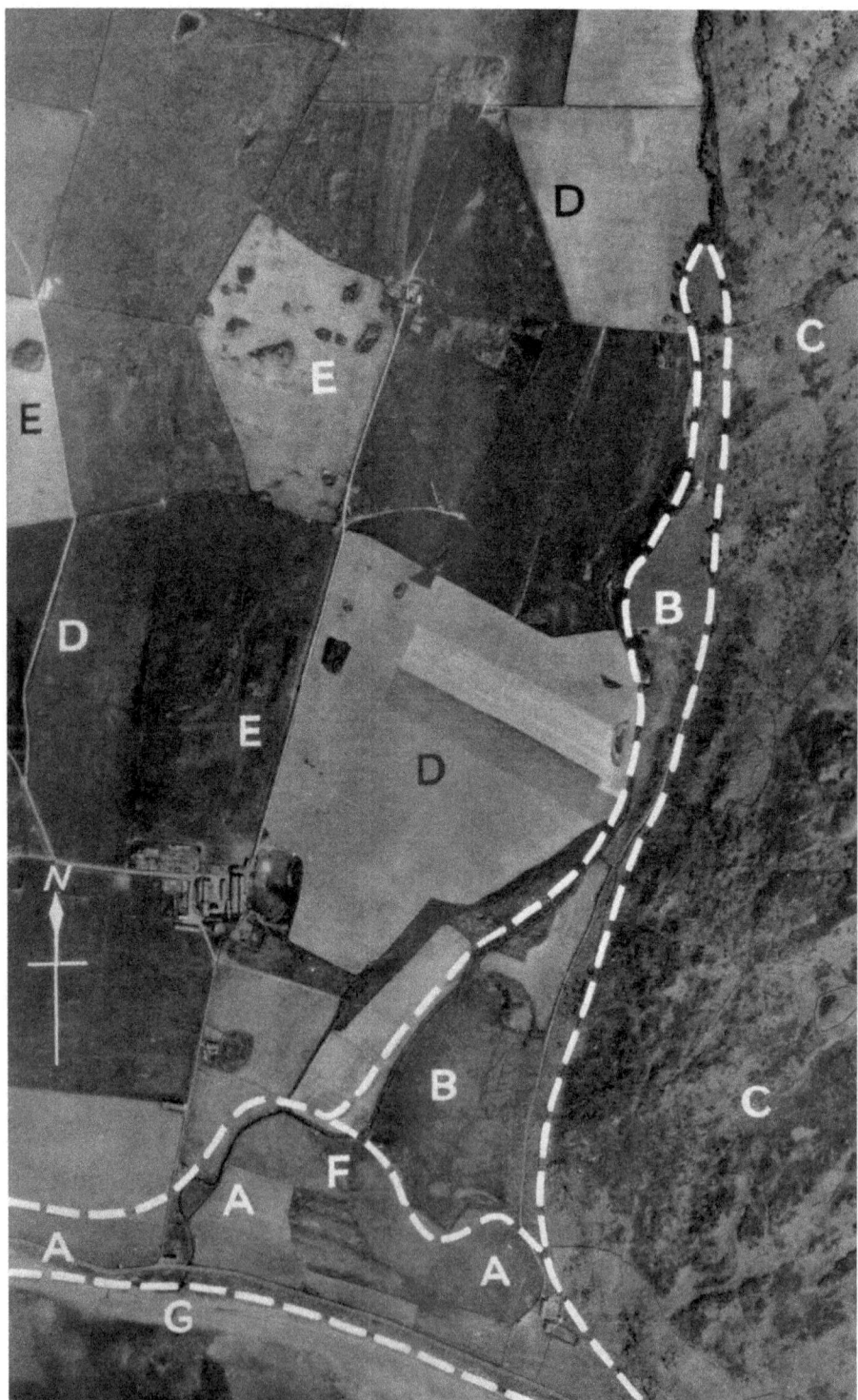

in the form of Epidiorite sills, and these rocks tend to form harder areas which stand up as ridges separating the lower zones which in most cases lie on the Quartzite. These hard and soft rock zones alternate along the coast in very narrow belts with the result that the landscape takes on a furrowed appearance which can be appreciated on the plate, particularly in photograph *B*.

Both photographs show the characteristic off-shore islands with long, narrow outlines, marking ridge areas now detached, and in particular Eilean Dubh is a fairly typical longitudinal coast island with its own detached longitudinal islets at the Northern end. The long, narrow strip of land marked *C* on photograph *A* is technically a peninsula since it is linked to the mainland by a neck of sand, except under abnormal conditions, but it is not difficult to envisage its complete separation to form an island entirely typical of the so-called Dalmatian type of coast. Where the outer ridges have been eroded the sea may break through to form bays, sometimes partially land-locked by the remains of the outer ridges. The bay in photograph *A* is not a good example, since only the promontory at *D* really shows this effect, but in Carsaig Bay in photograph *B*, especially on its Northern side, the promontories *E*, *F*, *G*, *H* separate small side bays, developed between the ridges, which give the whole bay its distinctive shape. The photographs were taken at high tide and therefore the sea covers most of the beach deposits, but nevertheless it is possible to see traces of the development of bay-head beaches near *G* and *D* and at *J*.

The only structural features which do not conform to the main Northeast to South-west trend lines are the Dolerite dykes which cross the area in a variety of directions between East–West and North-west–South-east, two of which can be detected on photograph *A* between *K* and *L* and between *M* and *N*. On photograph *B* a similar dyke can be traced from *O* to *P*, where its erosion has produced a shallow valley, and it will be noticed that a continuation of this line corresponds with the rather unusual direction of the sound separating Carsaig Island from the mainland. The evidence of the photograph would seem to suggest that the combination of weathering and erosion, ice action and marine erosion along the line of this dyke may possibly have been responsible for the development of this particular coast feature.

The 25-foot raised beach exists round most parts of this stretch of coast, but is rarely more than a narrow ledge in the area shown on the photographs except round Carsaig Bay, where its limits are marked by a dotted line and where it forms one of the few level tracts of the whole area which can be used for agriculture. Here the farms have the appearance and siting of typical crofter settlements, but it should be pointed out that in this case ease of communications between Carsaig Bay and the near-by larger settlement of Tayvallich means that here there is a great deal less isolation than in many true crofting areas.

Plate 58, by contrast with the preceding photographs, shows an area of transverse coastline in the South-eastern corner of the island of Islay just to the North of Ardmore Point. Here the separation of the island from Jura by the North-north-west to South-south-east Sound of Islay means that the coast has been developed nearly at right-angles to the North-east–South-west Caledonian trend lines continued from the mainland in the areas shown on Plate 57. Here the main rocks involved are the Port Ellen Phyllites and Epidiorite which occur in alternating narrow belts running in a generally North-east–South-west direction across the tip of the island. In this precise locality round Aros Bay (A) the calcareous Phyllites contain unusually great amounts of limestone and are abnormally soft, so that marine erosion has led to the development of the large bay of Aros cut into the end of this belt of Phyllite, which is marked inland by the wide valley through which the Kintour River (B) meanders towards the beach at the head of the bay.

On either side of Aros Bay the Epidiorite forms belts of much harder rock which produce areas of greater elevation on the flanks of the valley inland and have resisted marine erosion to form the promontories at C and D. In the latter case it is possible to trace the actual ridges produced by the Epidiorite and to relate these to the minor headlands projecting out to sea and, indeed, to the submarine rock ridges which the extreme stillness of the sea at the time of photography makes it possible to identify. Immediately North of Aros Bay the trend lines of the Epidiorite diverge from the general North-east–South-west direction and the consequent convergence of the lines of the promontories, small islands and submerged rocks near C, E and F serves to emphasize the effects of the structures on the coastal outline in such an area.

The photograph was taken near low tide and therefore gives a good impression of the distribution of beach material on such a coast. Near G, on a relatively gently shelving area between high-water mark and low-water mark, the beach material can be seen among the rock ridges below the true bay-head beach at G and the slightly developed bay-side beach above F. Near E, however, the steeper slope of the coast means that beach development is relatively slight and most of the material is under water even at low tide, though it can be seen as the light-toned area (H) among the submerged rocks.

Plate 59 has been arranged for stereoscopic viewing from three adjacent prints so that the details of marine erosion can be studied. In this case an area has been selected where geological conditions are comparatively uniform and therefore the processes of marine erosion have not been greatly modified by the appearance of different rock types on the coast. It shows a small stretch of coast immediately East of Dunnet Head where the gently sloping plateau of Caithness is composed of the Thurso Flags of the Middle Old Red Sandstone, overlain by a uniform sheet of Boulder Clay. Off-shore the water is deep, the 10-fathom line reaching as far inshore as the stacks at A, so that the waves usually remain unbroken until they reach the stacks and promontories of the coastline itself, and the area therefore presents a quite simple example of wave attack on a relatively straight stretch of rocky coast.

The sea cliffs are steep and increase in height from 50 feet in the East at B to near 100 feet in the West at C and this, combined with the deep water inshore, means that the erosional effect is confined to a very narrow belt near the foot of the cliffs. Between the stacks and the mainland, below B, and below D, there is evidence of the development of a wave-cut platform at about high-water mark, and along the sides of the bay of Kerry Goe (E) the same feature can just be traced. The present photograph was taken near low tide and therefore the platform is exposed, and the white appearance of surf, marking the areas of wave attack at this time, demonstrates the intensity of the erosional forces on the minor promontories, between C and D for example, leading to the development of stacks or submerged rocks (A, F). It should perhaps be explained that the difficulty of studying the sea margins of the rocks and stacks under the stereoscope arises from the movement of the water and spray between the two photographic exposures, which prevents the true fusion of the two images.

Beach development is characteristically lacking in this area of erosional coast forms, the only examples being the very small bay-head beaches at C and E in typically sheltered positions.

Wave attack is a phenomenon produced by movements which can be recorded photographically, and the very remarkable photograph A on Plate 60 is an instance where some of the mechanisms associated with this process can be illustrated more satisfactorily by such means than by any other technique. It demonstrates the case where the headland at Scarborough is being subjected to wave attack, whereas in contrast the wave energy is being largely dissipated in the wide bays on either side of this feature. The true sea waves may be seen as cloudy lines running diagonally across the photograph, but it will be noticed that the phenomenon of wave refraction results in these lines being curved round by frictional drag on either side of the headland at C and D so that wave attack is concentrated on all sides of the headland. One consequence of this development of curves is that the breaking waves round the headland are concentric on the outer curve of the projecting land and form backwash ripples which extend far out to sea. Indeed, one of the most remarkable features of this particular photograph is the fact that it is possible to trace these concentric ripples, superimposed upon the true waves, right to the seaward edge of the photograph, their curve being indicated by the dotted line EE_1.

In the bays at F and G slightly shallower water leads to the curving of the wave lines by this same phenomenon of refraction, and at G it is possible to see premature breaking of the waves over the shallows which is producing a minor group of translation waves with much higher frequency than the main wave system. Near G it is also possible to see the increasing sharpness of the line of the waves caused by their departure from true wave form as the front of the wave rises and curves over before breaking along the line of the beach.

The last photograph demonstrating the effects of coast erosion (photograph B, Plate 60) is chosen as a contrast to show an area where the rapid erosion of comparatively soft and uniform rocks has produced a long straight line of sea cliffs which are now cutting across relief features produced by normal land erosion. The area is on the South coast immediately to the East of Sidmouth where a number of well-developed valleys terminate abruptly in the steep sea cliffs. Here the cliffs are composed of the Keuper Marls, and this formation is also exposed along the floor of the valleys. The valley sides, however, are in the Greensand, whilst Chalk covers the remaining areas outside the valleys, though it is everywhere overlain by Clay-with-Flints except for a tiny area near C. The normal erosion of the Chalk and Greensand produced characteristic wide and rounded valleys whose lower portions have now disappeared as a result of the rapid backward erosion of the comparatively soft Keuper Marl cliffs, the smaller valley D being strikingly similar to the head of the larger valley at E and therefore possibly indicates the destruction of a considerable tract below this point. The abrupt termination of the larger valley by the sea cliffs makes it 'hang' over the beach, except in its lowest portion where an interesting small notch (F) appears in the face of the cliffs reaching down to the back of the beach. Such a feature could presumably develop as a result of backward cutting in a stream rejuvenated by the destruction of its lower course, but an alternative explanation is suggested by the existence of the chines of Bournemouth and the bunnies of Christchurch farther East. These features are trenched central portions of open rounded valleys which reach the sea in Bournemouth and immediately to

the East of it, and it has been suggested[1] that marine erosion might ultimately lead to the destruction of the chines. It therefore seems possible,
n view of the considerable degree of recession of the cliffs which has
already taken place to the East of Sidmouth, that the notch at F represents
the upper limit of a former chine or bunny.

Ample eroded material for beach development is available in this
locality, but in its present stage of development the beach in such a coastal
area is merely a narrow belt of material in transit, a good deal of which is
subsequently moved along the unbroken coastline by the combined forces
of wave action and the tides and currents.

When the total result of wave action, tides and currents leads to the
deposition of eroded material, the form taken by the deposits is infinitely
varied and is obviously governed in part by the configuration of the coast
as well as by the sea movements mentioned above. Some examples of
beaches formed in bays and on relatively open stretches of coast have
already been given, but the application of air photography to the study of
marine deposition is perhaps of greatest use where it can demonstrate the
development of the more unusual forms of deposition, such as spits and
bay bars, and where it may indicate the stages by which such land forms
modify long stretches of coastline.

Plate 61 shows the results of the closing of an old sea bay, marked now
by the line of the 25-foot raised beach round the inland side of Loch
Strathbeg on the North-east coast of Scotland to the North of Peterhead.
The Southward and South-eastward run of the tides on this part of the
Scottish coast, combined with a Southward movement of material due to
wave action, led to the development of a spit, growing in a generally
South-eastward direction from the Northern end of the bay under the lee
of a Boulder Clay cliff. Throughout the greater part of its length the spit is
composed of shingle derived from the rocky coast to the North round Kin-
naird Head and its extension Southwards converted the sea bay into a long,
narrow inlet entering the sea about $1\frac{1}{4}$ miles to the North-east of Rattray
Head (B). The inlet formed a natural harbour for the small town of
Rattray, the site of which is now marked by the buildings at A, and prior
to 1720 this harbour had quite a thriving trade. At that date, however, the
entrance to the harbour was finally blocked, not so much by the comple-
tion of the Southward growth of the spit across the bay but by the development of an area of blown sand at its Southern extremity, derived from the
beaches on the spit, which apparently sealed up the entrance very rapidly
indeed. This process therefore resulted in the sea inlet becoming the present
Loch Strathbeg, the town and port of Rattray virtually disappeared and

[1] Steers, *Coastlines of England and Wales* (Cambridge, 1948), p. 291.

the outline of the coast as far as Rattray Head was smoothed by the continued development of blown sand right out to the Head itself. The true spit is a wide one and consists of two main shingle ridges, the inner one overlooking the Loch being known as the Back Bar, and the outer one forming a crest immediately behind the present beach. At C, near the Northern end of the spit, the area between the ridges is marked by a well-defined 'low' which is partially filled with water draining from the loch. The present level of the loch is about 8 feet O.D. and when its Southern end was closed it began to drain Northwards through the shingle, a process which is now stabilized by the excavation of artificial channels (D). At EE_1 the shingle of the main spit is replaced by the blown sand which completed the closure of the bay, and the distinction between the two materials is extremely clearly shown on the photograph. This is partly due to the white appearance of the sand on the photographic print, but it is mainly caused by the very different morphology of the blown sand areas when compared with the regular rounded ridges of the shingle. In the sand area the development of a narrow zone of dunes has been replaced by a period of dune erosion which has led to the creation of blow-out gaps in this ridge, and it is possible to trace various stages in the evolution of these erosion features between the small and incomplete example at F and the wide gap through the dunes at G. The impressive height of the dunes in this area may be deduced from the length of the shadows cast into the gaps through them, those at G being three times as long as the shadows of the small house at H. The gaps through the dunes are all aligned with the dominant Northerly winds of this coast, and it is of interest to notice that where the direction of the coast changes at Rattray Head, though blown sand is present, it has not led to the development of dunes.

Plate 62 is intended to show the development of two further spits, each of which presents unusual features. Photograph A shows the two spits at the entrance of Pagham Harbour to the North-east of Selsea Bill which appear to have grown from opposite directions and have nearly closed the entrance. Apparently in this area a drift of beach material towards the North-east is known to exist[1] and is assumed to have led to the building of the spit (C) at the Western side of the harbour, but on the other hand there is no real evidence of an opposed drift towards the South-west which might explain the growth of the longer spit on the other side. It is of interest to notice that on the photograph this anomalous state of affairs is confirmed by the fact that the wave direction at C is such that one might expect wave transference of material along the line of the spit,

[1] Steers, op. cit., pp. 304-5.

whereas at D the wave movement is directly on-shore at the time of photography. It is possible, of course, that the study of photographs taken under widely different conditions of tide and wind direction might afford clues to the explanation of the origin of the second spit.

The extremely broken appearance of the water between the ends of the spits indicates the existence of a current or scour through the constricted harbour entrance—a feature which often explains the failure of spits to complete the process of closing off bays and estuaries in a fashion comparable to that shown on Plate 61. The slight recurve of the Eastern spit and the development of a lobe at its termination may be linked with the presence of this current and the on-shore waves, and it will be noticed that there is an irregular area of deposition near the end of the shorter spit where the current through the entrance is causing the main waves to break at E.

A point of general interest on this print is the variation in the photographic tone on the water, ranging from black on the enclosed area at F, through the darker greys on the comparatively still water in the harbour at G, to a light grey on the sea and in the harbour entrance where the surface texture is obviously appreciably more irregular.

Photograph B shows the unusual double spit of Dawlish Warren at the mouth of the Exe as it appeared in 1946. This spit is now apparently in the process of being destroyed by erosion of its outer edge, but for a long period it consisted of an inner Warren (C) and an outer Warren (of which traces can still be seen at DD), separated by a continuous low (E).[1] By the time of the 1946 photograph the erosion had already largely removed the outer spit, and at the present time the inner warren itself is subject to intermittent wave attack.

The initial development of the spit by the Eastward drift of material has been attributed largely to wave action, and the direction of wave approach shown on the photograph indicates how normal wave transference of material may have brought this about. On the other hand it is now suggested that diminishing supplies of material has led to this same wave action eroding the outer edge of the Warrens and transferring the original material still farther East.

Another distinctive feature of the spit in 1946 was the marked development of a recurved end where the deposits were carried round by current and wave action into the entrance to the Exe estuary, some of the material no doubt being derived from the main part of the outer warren.

Where changes of the relative levels of land and sea have resulted in an

[1] See C. Kidson, 'Dawlish Warren', in *Institute of British Geographers. Transactions and Papers* (1950), pp. 69–80.

apparent raising of the land level the effects of marine erosion and deposition are seen away from the present coastline in the form of raised beaches, and the two remaining photographs of coastal areas are devoted to two totally dissimilar examples of this phenomenon. Plate 63 shows a portion of the Ayrshire coast North of Ballantrae and immediately South of Bennane Head. Here the 25-foot raised beach (A) lies between the present shore and an impressive line of inland cliffs which were the sea cliffs of the former coastline. The change in sea level, relative to the land, has meant that the base level of erosion of the streams has been lowered so that the streams C and D have been rejuvenated and are excavating deepened valleys inland, the valley at C showing remarkably good examples of the resultant terracing.

Though the geology of the area shown on the print is extremely complex, most of the details are hidden under a drift of Boulder Clay and moraine, but the unusual appearance of the surface round E is due to the presence of a considerable tract of Spilitic Lava in that locality which is not completely hidden by the superficial deposits.

The raised beach shown on Plate 64 is not backed by former sea cliffs since it occurs on the low-lying coastal area to the East of Luce Bay in South Scotland, at the mouth of the Gillespie Burn. The raised beach is outlined at AA and it will be seen that it separates the present narrow beach from the area of river alluvium (BB) extending up the narrow valley of the burn. It would appear that the deposition of beach material has led to the diversion of the Gillespie Burn on two and possibly three occasions, since it will be seen that the stream turns abruptly to the right at the back of the older beach at F and at the back of the present beach at G, whilst an intermediate beach stage may be represented by the deflection of the stream midway between these points.

The remarkable contrast on the photograph between the area CC to the East of the burn and DD to the West of it is a product of ice action. In the area CC the Silurian rocks have been ice scraped and have a very thin patchy soil cover and a sparse moorland vegetation, whereas the area DD is almost completely covered with Boulder Clay, except where the rock appears at EEE, and is in use for mixed farming.

Chapter 6

HUMAN GEOGRAPHY

It is obvious that almost every aspect of human activity can be recorded in some form on air photographs and in so far as the need arises it is possible to study the results of this activity with great precision on large-scale photographs. Exact information about such minute details as building shapes and types, garden patterns, farming practice, the plan and function of industrial undertakings or the design and lay-out of railway and road systems can be determined, sometimes with greater facility than by studies in the field. Perhaps the best use of this method, however, is to be seen where it makes possible a clearer appreciation of the interrelations which exist between the physical landscape and man's activities, and most of the examples below have been selected to that end.

THE RELATIONSHIPS BETWEEN RELIEF, SOIL TYPES AND SETTLEMENTS

Valley Settlements

In the foregoing sections a number of examples of the distribution of settlement in relation to various types of river valleys have already been discussed (Plates 14, 19, 34, 37), and the present group of photographs has therefore been selected in order to include further examples which have additional points of more general interest. Plate 65 shows three of a long line of so-called 'by' settlements along the Eastern edge of the valley of the Ancholme in Lincolnshire. The villages are (A) Grasby, (B) Owmby, (C) Searby, all of them names of Danish origin associated with the arrival of Danish elements in this area along the line of the River Ancholme from the Humber. In the case of Owmby and Searby the first element in the name is a Scandinavian personal name, but in Grasby it is a descriptive element meaning 'a stony place'.[1] This is appropriate when one recalls the line of approach along the river alluvium and Kimmeridge Clay of the valley since on arrival at Grasby the settlers came to the edge of the Lincoln Wolds where the Red Chalk and Lower Chalk outcrop, and, after the softer materials of the valley, the presence of Chalk no doubt made the stony character of the locality of Grasby the most remarkable feature. The edge of the Chalk has been marked on the print as the line *DDD*, but the remarkably abrupt change of tone on the print makes the junction

[1] Ekwall, op. cit., 'Grasby'.

abundantly clear. It will be seen from the form lines that there is a scarp-like edge with a height of some 150 feet along the line of junction of the Chalk with the Kimmeridge Clay which it immediately succeeds in this area and which, in this precise locality, is not masked by superficial deposits of sand. The distinction between the Chalk and Clay on the photograph is partly a function of their different soil colours, but it is also a measure of the much greater extent of grass farming on the Clay when compared with the exclusively arable use of the Chalk. This agricultural distinction points to one of the probable reasons for the choice of site of the village line, since here settlements were well placed to take advantage of the different agricultural opportunities of the two soil regions, and it is of interest to notice that locally the Wolds are referred to as 'tops' bearing the names of the villages below, from which they were presumably cultivated, such as Owmby Top and Grasby Top for example.

The villages, lying on the lower slopes of the scarp, are at, or immediately below, the spring line at the junction of the Chalk and the Clay and there-fore enjoyed an adequate water supply, whereas their elevation ensured that they were well clear of the ill-drained land and the liability of floods on the floor of the valley.

It will be noticed that, as in the case of the settlements on the Lincoln Edge, the villages do not lie on the line of the road following the crest of the scarp, a distribution which bears out their origin as settlements based on movement from the valley rather than along any earlier routeway which may have existed along the Chalk upland.

Plate 66 shows valley settlement on a very much smaller scale which is typical, however, of many such sites in the Chalk country. The valley is that of the Winterbourne, an intermittent stream flowing Southwards across the Marlborough Downs near Avebury, and it produces little more than a shallow depression in the Chalk surface with a narrow belt of Valley Gravel in its lowest portion and an even narrower strip of Alluvium along the stream itself. Nevertheless, the existence of different soils from those of the surrounding plateau-like stretches of the Lower Chalk, and in particular the possibility of the development of meadows along the valley soils meant that this was a favourable site from an agricultural point of view. In addition, throughout the greater part of most years, a water supply was available either from the bourne itself or from springs and shallow wells, and this tended to lead to the slight nucleation of settlement in the village of Winterbourne Monkton (A) which contrasts with the relatively sparse settlement of the Lower Chalk areas with their virtual absence of surface water.

The actual form of the settlement of Winterbourne Monkton is interesting and quite characteristic of the smaller valley settlements in this part of the Chalk downs. It consists of three fairly large farms and a very limited number of small houses and cottages and is obviously a purely agricultural grouping from which the surrounding arable land is worked, whilst stock can be kept for certain parts of the year on the grass of the valley, conveniently near to the farms themselves. It will be noticed that the buildings are grouped in two areas a little way back from the stream, a feature which is much more marked in the larger settlements of the Avon valley to the East, where the danger of flooding is appreciably greater.

Apart from the valley, the nearly level surface of the Chalk is broken in this locality by two slightly higher areas of very gently rounded relief where the Middle Chalk overlies the Lower Chalk, the outcrop of the former being marked at B and C. In the area B the Middle Chalk forms Windmill Hill, on the crest of which the photograph shows clearly the nearly concentric ditches and the 'tumuli' of the famous Neolithic Windmill Hill site from which a major part of the typology of the Neo-lithic cultures of Britain was derived. At C the photograph just includes the Northern part of the curious outer bank and ditch surrounding the megaliths and stone circles in the village of Avebury, which is one of the most important prehistoric sites of the Chalk downland.

Turning now from purely agricultural communities to an example of a valley where a moderate degree of industrialization has taken place, the contrasting distribution of settlements is immediately apparent. The physical geography and geology of the area shown on Plate 67 has already been discussed in connection with Plate 25 above, and it will be sufficient to recall that the lines marked on the print represent the outcrop lines of the Lower Oolite on the hills, the Upper Lias Sands on the sloping valley sides, and the Upper Lias Clays on the valley floor. Here, valley settlement has been very closely linked with the great development of the West of England woollen industry in the valleys converging on Stroud and it will be seen that even in this short stretch of the valley three factories still remain as evidence of its former importance, though throughout the Stroud region many of the woollen mills have been converted for the manufacture of light engineering, furniture and plastics products, whilst others have maintained their connection with the wool trade by turning to the manufacture of high-quality specialized textile goods. It will be seen that the three factories all lie on the actual line of the stream course and each has diverted a mill-race through the works itself in order to use the water for power and for processing. The development of the woollen industry, of course, ante-dated canal and railway construction, and it is of interest to notice that the mills have no direct link with the development of trunk communications through the valley. The canal is the Thames and Severn Canal and the railway is the old Great Western line linking the Stroud area and the railways of the Vale of Severn with Swindon and London, but though these communications did serve as a long-distance link for the whole of this region, the mills themselves were served entirely by the road which forms the boundary of their sites on the East.

On the valley floor non-industrial settlement is almost entirely absent or is restricted to a small number of houses built close to the factories along the road mentioned above. This was partly due to the fact that the only entirely suitable sites in the valley had been given over to industrial buildings, but in addition it was no doubt also due to the ill-drained character of the clays and the coldness and dampness of the lowest parts of these deep valleys. The villages of Thrupp (F) and Rodborough (G) are therefore strung out as long lines of houses along the roads which follow the contours of the middle slopes of the valley on the Upper Lias Sands. Here there were excellent dry building sites and a position clear of the less favourable conditions of the valley floor, whilst the light soils have favoured the development of gardens and small-holdings, particularly on the Eastern side of the valley, at H for example, with its good slope and aspect. Elsewhere the landscape, especially on the Oolite, takes on the park-like appearance characteristic of the central Cotswolds, with considerable numbers of large farms and estates.

Where valley settlements are in the form of towns the factors governing their site, development and present characteristics are so varied that obviously no generalizations can be arrived at, and therefore the three succeeding plates have been chosen to illustrate cases where it is the presence of the valley itself which has been the main factor influencing the settlement pattern.

Plate 68 shows the town of High Wycombe which is often quoted as an example of a gap town, since it lies in the valley of the River Wye where that stream has cut a deep trench-like section through the Chalk of the Chilterns before joining the Thames at Bourne End above Maidenhead. This gap along the valley of the Wye has been of importance since the Roman period when it apparently provided the route of a minor Roman road from the line of movement along the Icknield Way near Princes Risborough (see below, Plate 76) to the Thames valley, though even earlier importance may perhaps be inferred from the presence of hill forts on the Chalk summits on either side of the valley and finds of Bronze Age materials in the valley itself.

One interesting aspect of the photograph is that the oldest part of the town can easily be identified by the clustering of irregularly placed buildings in the elongated oval area immediately South of the marked curve of the railway line, where the present canalized line of the river (*A*) indicates the location of the original settlement in the centre of the river-valley gap. The development of this old town really only dates to the post-Conquest period when it was known as Chipping Wycombe— a name which signified its real character as a market town. In common with many such market centres of this period the town included a characteristically wide and straight market-place or street which today forms the main road of the town at *B*.

In more modern times High Wycombe has continued to function as a route centre, partly because this gap through the Chilterns is used by the main railway line from London to Banbury, Oxford and the North and West of England, and partly because the road through it now serves as one of the main lines of communication between London and the Thames valley and Oxford, Gloucester and the West of England and South Wales. Here too the main road (*CC*) along the dip slope of the Chilterns from Reading and Marlow to Amersham and Berkhamstead crosses the main North-west to South-east line of communications.

In addition, High Wycombe has become an important manufacturing centre and some industrial buildings can be seen among the smaller ones of the town, especially near the railway. On the other hand the constriction of the valley site has meant that the industrial development has taken place outside the old town, partly along the line of the main valley beyond *D* and *E*, but also along the tributary valley at *F*. Recent housing construction has also been compelled to spread in either direction along the valley and on the flanks of the tributary valley at *G*, and in these outer areas it is interesting to note the disposition of the houses in long rows in contrast with the irregular arrangement in the old town. In addition, housing has also spread on to the higher ground on the Chalk spurs along which the Reading–Amersham road (*CC*) approaches the valley, and in this case it will be noticed that it is almost entirely the larger and well-spaced houses which have been built along the line of this road. The latest phase of house construction is represented by the area *H* which, of course, represents a planned endeavour to overcome the difficulties of the constricted site of High Wycombe by spreading out on to the Chalk upland. The exceptionally wide roads of deliberately curved line with interspersed cul-de-sac streets are, of course, entirely typical of one variety of housing estate design.

Plate 69 shows the site of Stirling as an example of a river-crossing town which derived additional importance from the fact that it is situated in one of the few easily defended localities in an extremely flat lowland. The elaborate meanders of the Forth give evidence of the level character of this part of its valley which lies within the area of the 50-foot raised beach, but at this point the valley is crossed by an important belt of intrusive igneous material which has led to the development of isolated hill masses, portions of two of which are to be seen on the photograph at A and B. The strategic importance of these hill areas across the line of the valley will be immediately apparent and both have been used for defence. The remains of an ancient fort exist on the hill B, but as far as Stirling itself is concerned it is the site of the Castle (C), on the hill mass (A) which is of prime importance. The precise location of this commanding site meant that it could block the landward approaches to the narrow neck of land enclosed within the main meander which was protected on all other sides by the river itself. There was, therefore, available a well-protected area at a point at which bridging operations were practicable, and in these two facts there lies almost the whole explanation of Stirling's early development.

The subsequent building of bridges farther downstream may appear to have destroyed some of Stirling's significance, but the close juxta-position of road and railway bridges serves to demonstrate the continued importance of this route which still carries the greater part of the traffic both by road and rail from the whole of North Scotland to Glasgow and to England by the West coast route.

In the area within the meander and near the railway there are now mainly large public and administrative buildings, but on the slopes of the castle hill below D the irregular building arrangement of an old town can still be observed. As in the case of High Wycombe, the restriction of space imposed by a highly specialized site has meant that the limited industrial development of Stirling has taken place on the outskirts of the town at Forthbank (E) whilst modern housing has had to expand towards the East at F and towards the significantly named Causewayhead (G).

Although Stirling is not a large town it will be noticed that there is a fair-sized railway marshalling yard near the very fine station (H). This is rendered necessary by the considerable volume of traffic which passes through this bridge point but which has to be marshalled here according to destination along the four lines to the North or made up into main-line trains for the South.

Lewes (Plate 70) is not strictly speaking a valley settlement, but its development has been so closely linked with the valley of the Ouse that it may perhaps be included in this section. Here the Ouse has cut down through two parallel structures, the Falmer syncline, the line of which is now marked by the projecting belt of Chalk upland North of the Downs, and the Kingston–Beddingham anticline, the erosion of which has produced a valley of elevation which includes the low-lying land round the Ouse known as the Brooks (*AA* on the photograph).[1] The gap through the former feature therefore leads to a marked constriction of the valley of the Ouse between two low-lying open valley tracts, and the original site of Lewes was at this point on the spur on its right bank along the spine of which can be seen the High Street running downhill to the bridge. It is possible that there was a Saxon settlement on this site and supposed Saxon earthworks on the hill behind the town may mark attempts to defend the approaches from the West along the Chalk, a function later carried out by the Castle (*B*) which also had the rôle of controlling the valley route through the river gap. On the river side, Lewes was protected by the Ouse itself and its right-bank tributary, and in this defensible setting it developed from the tenth century onwards as a market town and a port, since it only lies about 8 miles upstream from Seaford Bay.

In later times Lewes continued to act as a market centre, but in addition the convergence of routes on the gap of the river valley has made it an important communications centre throughout its history. From the North main roads converge from the whole of South-eastern and Central England and from London, though it is interesting to notice that they converge well to the North of the town and approach Lewes as two roads following the edge of the Chalk above the valley bottom, the bridge in the town acting as a link between them. To the South, two roads (*C* and *D*) again keeping to the Chalk round the edges of the Brooks, lead to Eastbourne, Hastings and Dover and to Newhaven respectively, whilst the continuation of the High Street to the West leads to Brighton and the South coast.

Railway communications also converge on the town, three lines to the North leading in the direction of London, one of them using a tunnel under the town, whilst to the South a short line leads to Newhaven and two other lines turn East and West to follow the South coast to Eastbourne, Hastings and Dover and to Brighton, Portsmouth and Southampton respectively, the transfer of traffic between these lines necessitating the elaborate system of railway curves appearing on the photograph.

There is very little evidence of industrial activity on the photograph apart from the large number of quarries in the Chalk to the East of the river and the cement works immediately North of *C*. It should be pointed out,

[1] Wooldridge and Linton, 'Structure, surface and drainage in South-east England', in *Institute of British Geographers, Transactions* (1939).

PLATES 65—80

Plate	Locality	Ordnance Survey Sheet no. 1:63,360	Latitude and Longitude	Sortie number	Print numbers
65	Grasby near Caistor	104	W 00°22' N 53°32'	CPE UK 2563	3100 3102
66	Winterbourne Monkton	157	W 01°51' N 51°26'	CPE UK 1821	1082
67	Golden Valley near Stroud	156	W 02°12' N 51°44'	58/503	5035 5037
68	High Wycombe	159	W 00°45' N 51°37'30"	CPE UK 1965	4142 4144
69	Stirling	Scotland 67	W 03°56' N 56°07'30"	106G Scot UK 93	3354 3356 4354 4356
70	Lewes	183	E 00°01' N 50°52'30"	541/504	3149 3150 4149 4150
71	Dartmoor near South Zeal	175	W 03°55' N 50°43'	CPE UK 2491	3366
72	Claverham Somerset	155	W 02°48' N 51°23'30"	CPE UK 1869	4139
73	Old Sodbury Gloucestershire	156	W 02°21' N 51°32'30"	106G UK 1416	3381
74	Sutton Valence near Maidstone	172	E 00°36' N 51°13'	541/536	3213 4213
75A	Queen Charlton near Bristol	156	W 02°31'30" N 51°24'	CPE UK 1869	4172
75B	Barsolis Hill South Scotland	Scotland 88	W 03°57' N 55°00'30"	106G Scot UK 52	3222
76	Icknield Way Chinnor	159	W 00°53' N 51°42'30"	106G UK 1379	4109 4111
77	Fosseway near Chippenham	156	W 02°13' N 51°31'30"	106G UK 1416	3231
78A	Cerne Abbas near Dorchester	178	W 02°29' N 50°48'30"	CPE UK 1974	2377
78B	Ranscombe Bottom near Calne	157	W 01°57' N 51°25'	106G UK 1415	4068
79	Near Kelso	Scotland 81	W 02°32' N 55°35'	CPE Scot 315	3168
80	Winford near Bristol	165	W 02°39' N 51°23'	CPE UK 1869	3137

Plate 65

Plate 66

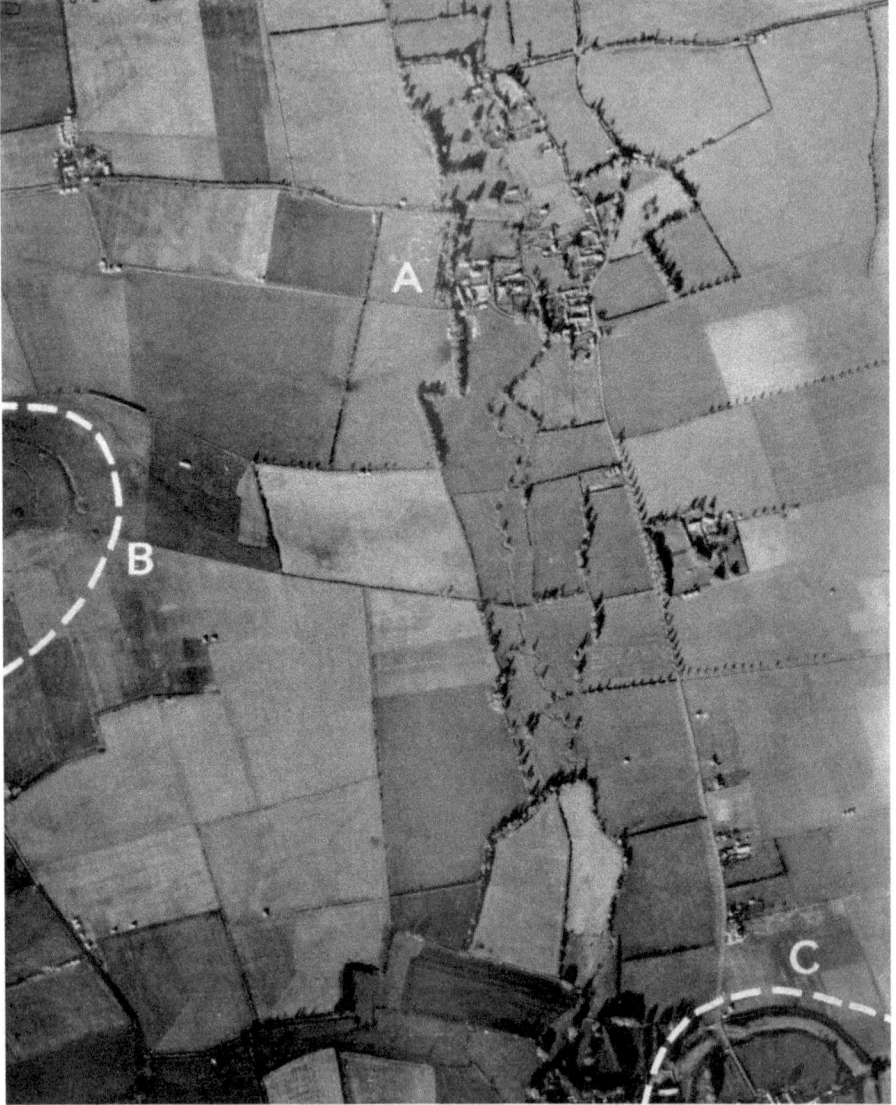

Plate 67

Plate 68

Plate 69

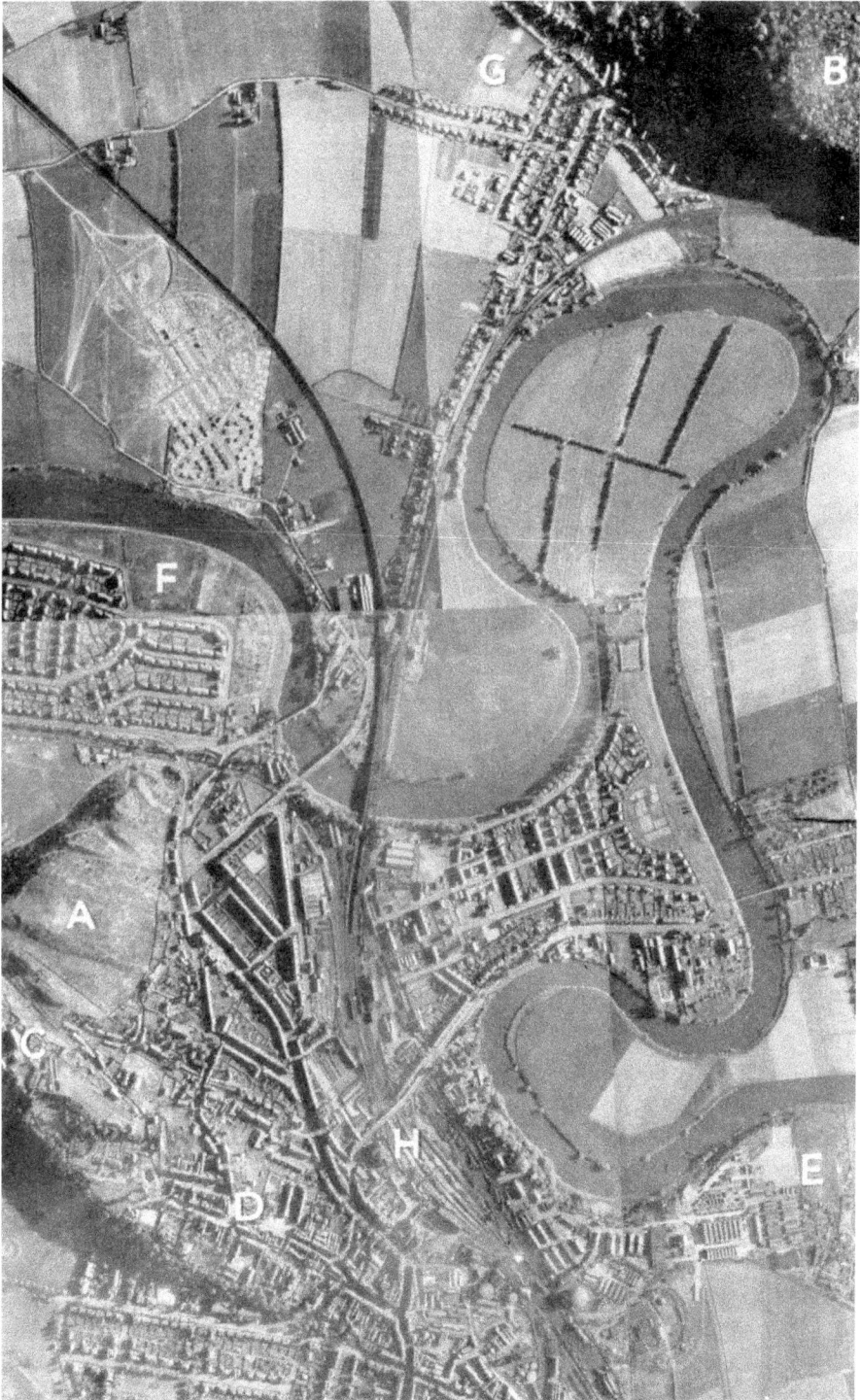

Plate 70

Plate 71

Plate 72

Plate 73

Plate 74

Plate 75

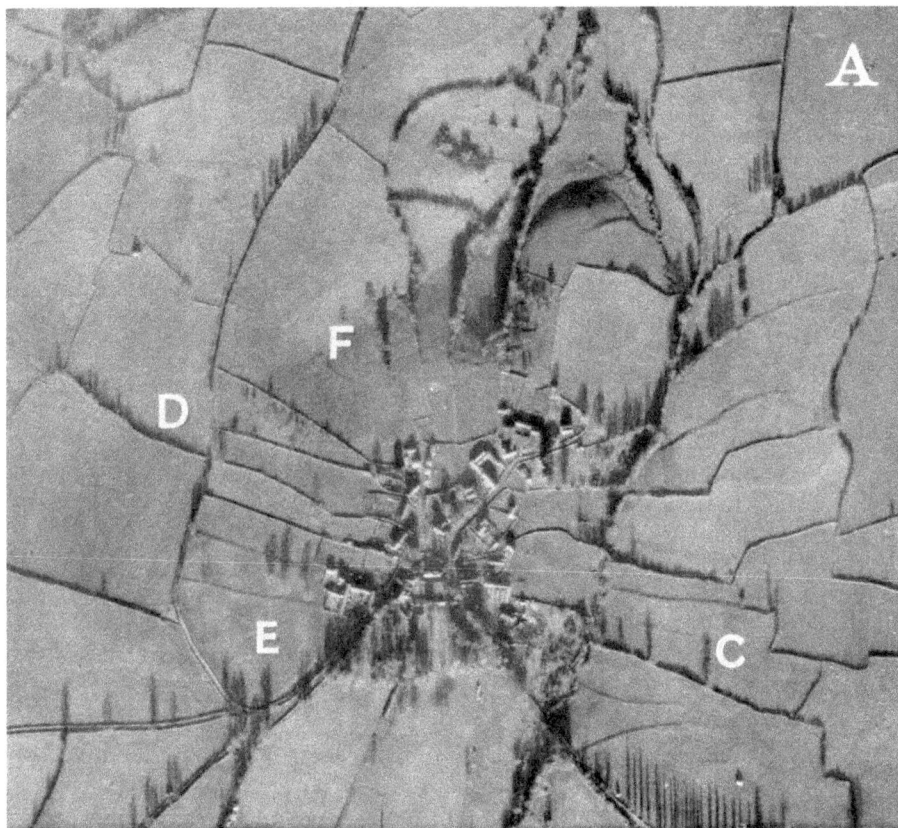

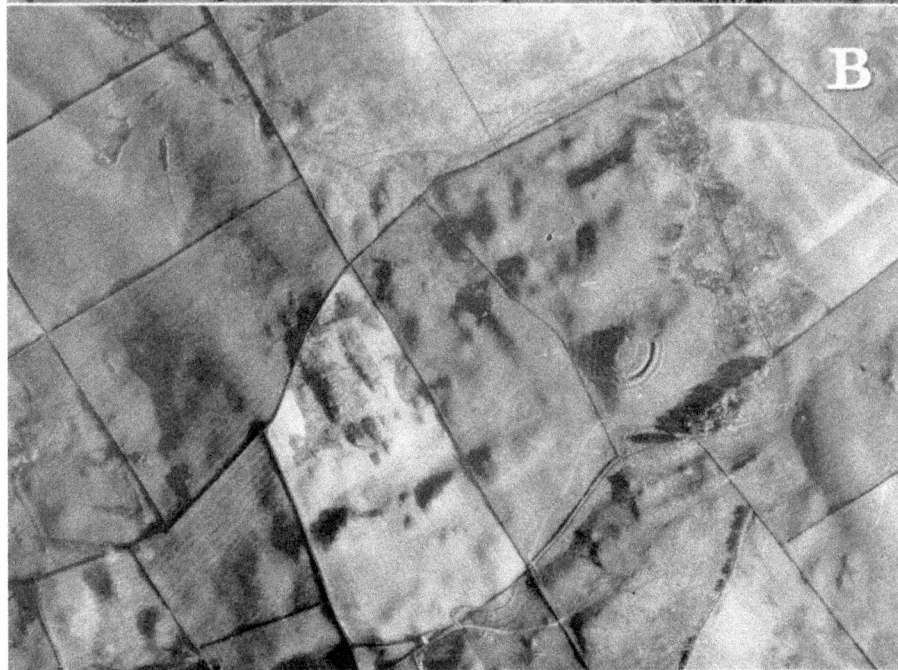

Plate 76

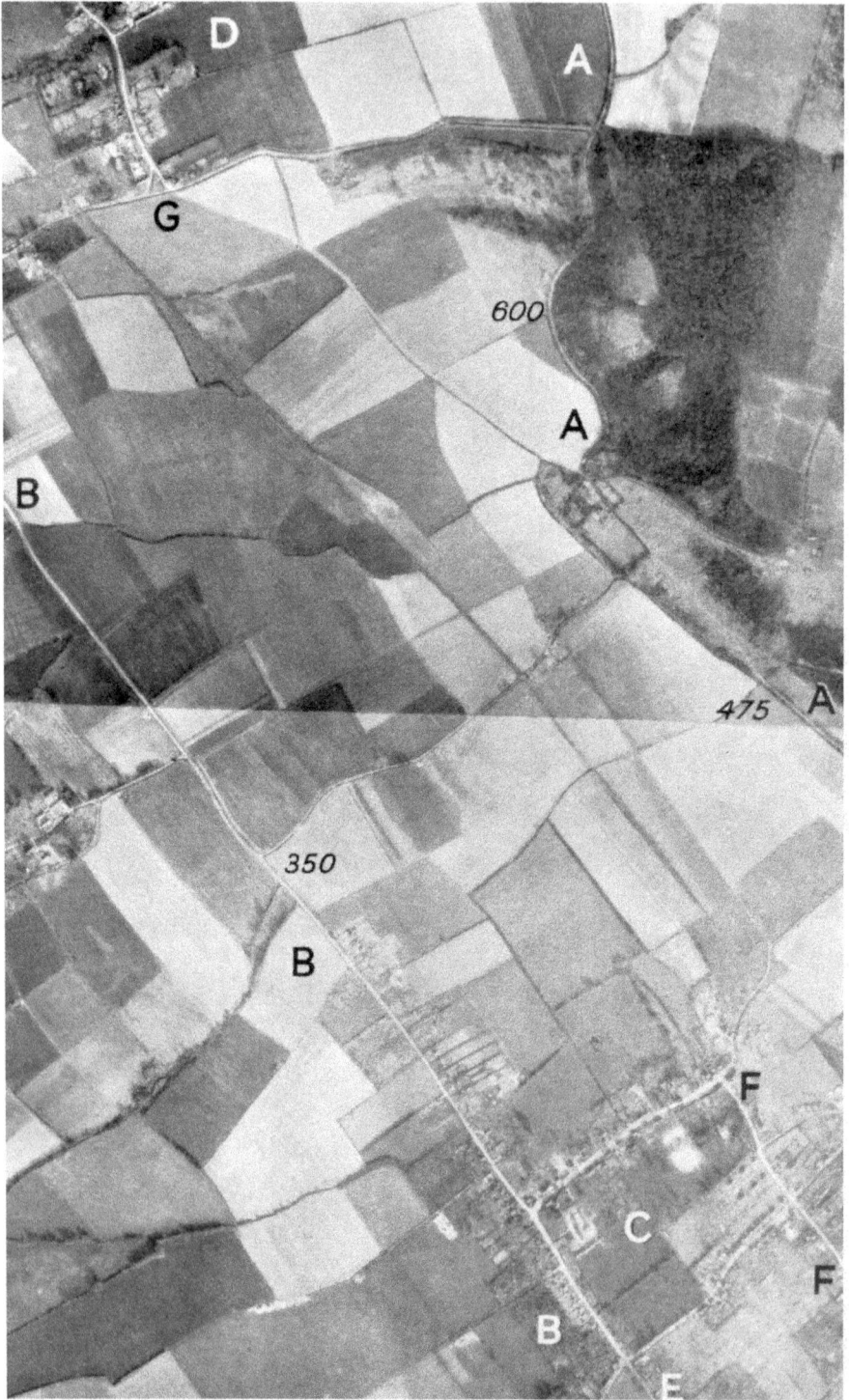

Plate 77

Plate 78

Plate 79

Plate 80

however, that the present photograph only shows the older central portion of Lewes and that the greater part of the modern development of the town, especially from a residential point of view, has taken place in a Westerly direction on to the Chalk upland where there was available open building land.

Non-nucleated Settlements

The study of valley settlements has almost inevitably led to the foregoing plates being concerned exclusively with nucleated settlements, and it is therefore proposed to include at this point photographs to show the distribution of settlement in two areas where there are no marked nuclei in order that their contrasting cultural landscape may be appreciated. The physical environments of the two areas are totally dissimilar, one being in the plain of Somerset and the other on the edge of Dartmoor, but though their landscapes are totally different it will nevertheless be seen that from the point of view of settlement distribution (but not density) they have a number of features in common.

Plate 71 shows the North-eastern flank of Dartmoor immediately to the South of the village of South Zeal, the greater part of the area shown lying on the Granite of Dartmoor itself, though a small area of the surrounding lowland is seen in the top right-hand corner of the photograph. In the bottom left-hand corner of the photograph the uncultivated higher parts of Dartmoor have a characteristic moorland vegetation and are unenclosed, but below this area the slopes of the moor have been enclosed into a patchwork of small fields or intakes, though it is apparent from the photograph that the majority of these are now little more than enclosed areas of rough grazing. Nevertheless, the enclosure clearly indicates that this area has been in agricultural use, but it is perhaps indicative of the poverty of its resources that there is no village or hamlet in this belt and the nearest nucleated village is South Zeal on the plain to the North. Instead, the settlement here consists of isolated farmsteads or even cottages, some of which are merely outlying buildings attached to larger but equally isolated farms on the surrounding plain. Such a distribution is fairly common on the flanks of the moorland areas in the South-west peninsula, throughout Wales and in Southern Scotland, and contrasts sharply with the pattern of settlement in the English plain.

On the other hand Plate 72 shows a portion of a relatively densely peopled, rich agricultural region round Claverham in North Somerset, about 10 miles South-west of Bristol. The collection of buildings round A in the centre of the print is known as the village of Claverham, but it is obvious from the photograph that here there is no real grouping of settlements into villages but rather a widespread distribution of individual houses and farms, the only apparent pattern being due to a tendency for these to be concentrated slightly along the irregular road pattern of the district.

Apart from a very small area of alluvium near the railway line in the top left-hand corner and a narrow strip of Dolomitic Conglomerate and Carboniferous Limestone in the area along the bottom edge of the print, the whole of the district lies on the Keuper Marls, affording reasonably rich soils of a uniform character which can be worked fairly easily. In addition there are no definite springs in this locality, but shallow wells in the Marl or Conglomerate produced supplies of water adequate for the limited demands of individual houses or farms, possibly as a result of water stored in the Limestone and Conglomerate to the South. Here then, in a district without marked relief features to single out a site for larger settlements, there was not the incentive to congregate round a single source of water supply whilst there was everywhere an adequate agricultural basis for the development of small farms, and the basic settlement pattern seen on the photograph seems a natural consequence.

In addition, however, it should be pointed out that the area lies between the main Bristol–Taunton railway line (B) and the Bristol to Weston-super-Mare road (C) at a distance of only 10 miles from the former city, a fact which accounts in part for the prosperity of farming here and for the existence of a number of purely residential settlements. The factory (D) does not represent an intrusive element into the rural landscape since it is one of a group of tanneries in this part of Somerset which draw some of their supplies of skins from the predominantly animal husbandry of this region.

Settlement Patterns in a Scarp and Vale Landscape

It has already proved necessary to describe photographs of various parts of the scarplands, for example Plate 25 in the Cotswolds, Plate 26 on Lincoln Edge, Plate 65 at the edge of the Lincolnshire Wolds and Plate 70 North of the Downs at Lewes. As the distribution of settlement shown on each of these photographs has already been discussed it is proposed now to include only two further examples to illustrate points not so far covered.

Plate 73 is a second example from the central Cotswold scarp, included because it is possible in this locality to demonstrate successive stages in the human utilization of the scarp belt. The oldest site shown on the photograph is Old Sodbury Camp (A) which is known to be of pre-Roman origin. It represents the prehistoric use of the open crest of the Jurassic scarp as a settlement area and as a line of communications probably forming part of the Jurassic Way across England, already referred to.[1] The road (E) is a portion of the present Bath–Cirencester road which, in this part of the Cotswolds, is probably along the line of the old trackway. It may also have functioned as a Roman route alternative to, or earlier than, the Fosseway, and it has been suggested that Old Sodbury Camp may have been modified and altered to its present nearly rectangular plan by the Romans. The camp is situated exactly at the crest of the scarp and it will be noticed that the impressive double ditch and rampart only exists to protect it from attack across the very gentle dip slope, there being a comparatively shallow ditch on the side overlooking the very steep scarp face.

Saxon settlements in this area are represented by the villages of Little Sodbury (B) and Old Sodbury (C) which form part of the long line of such sites on the Middle Lias Marlstone ledge near the foot of the Cotswold scarp from Bath to Wotton-under-Edge in the North. It will be noticed that at D, D_1, D_2 and D_3 there are traces of lynchets on the slope of the scarp immediately above the Marlstone ledge indicating the cultivation of this zone with which the original settlement sites were associated. Below the Marlstone ledge, on the other hand, the level land of the Lower Lias Clays (appearing at the edge of the print) was formerly ill-drained, and the siting of the villages on the ledge raised them above this less favourable zone but still enabled them to combine in their agriculture the use of the pastures and meadows on the Clays with the arable on the Oolites and Sands above.

It is open to doubt whether it was common for the cultivation of the actual crest of the scarp and the near-by portions of the dip slope to be carried out from the villages on the scarp face. Elsewhere along this part of the Cotswolds there are equally old settlements along the crest at Tresham, Hawkesbury Upton (Plate 92), Starveall and near the Cross Hands, and it is possible that isolated large farms, like the one which adjoins Old Sodbury Camp and a similar one immediately above Horton, have very ancient origins. One of the difficulties of the villages and farms on the crest of the scarp has been to obtain water, prior to the construction of a piped supply, and most of them depended on ponds or small wells and supplies carried from springs at the base of the Fuller's Earth. On the other hand the villages along the line of the Marlstone ledge are all situated near excellent springs and streams, and this may have been an additional factor in determining their site.

[1] See W. F. Grimes, 'The Jurassic Way across England', in *Aspects of Archaeology in Britain and Beyond*, Essays presented to O. G. S. Crawford.

Still later stages in the redistribution of settlement in this area led to an even greater departure from the line of the scarp with the development, some two miles farther out into the vale near the Limestone rim of the Gloucestershire coal-field, of the market town of Chipping Sodbury, a photograph of which is included later as Plate 86.

Plate 74 shows one of the less famous but nevertheless extremely interesting and important zones within the Weald, and it has been selected because it demonstrates the fact that certain features of the settlement pattern of the scarp-lands tend to recur even on features a good deal less impressive than the North and South Downs or the Chilterns for example. The area lies in the scarp-foot zone of the Weald on its Northern side, some 5 or 6 miles to the South-east of Maidstone at the village of Sutton Valence (A). The greater part of the area lies on the Lower Greensand but the dotted line at the right-hand side of the photograph marks the limit of this outcrop and its junction with the Weald Clay to the South. The edge of the Greensand forms a low Southward-facing scarp above this line as indicated by the spot heights at the top of the photograph. The character of the belt of Greensand in early times may perhaps be deduced from the great frequency of occurrence of the place-name element 'chart' in this area, derived apparently from Old English 'ceart', 'a rough common overrun with gorse, broom, bracken'.[1] Nevertheless, in contrast with the Weald Clays to the South, it was a generally favourable area for early settlement and its soil was capable of supporting arable agriculture so that from Saxon times, at any rate, the Lower Greensand belt was one of quite dense settlement. Unlike the scarp-land examples already studied, however, this settlement does not take the form of compact villages strung out along a single line but is rather a broad zone of smaller settlements which, in the area shown on the photograph, are aligned with two remarkably parallel roads following the Greensand outcrop. The left-hand road (BBB) almost certainly lies along the line of an ancient trackway known as the Chartway, and at C the small hamlet is still known as Chartway Street, whilst the names Chart Sutton and Chart Corner also occur on the road farther West. Apparently the Chartway is of pre-Roman origin since two minor roads from a Roman site in the neighbourhood of Maidstone ended abruptly along its line[2] and it may, therefore, represent another example of the importance of routes and settlements along the crests of the scarp-lands in prehistoric times. It is possible that the second road, almost along the crest of the actual Greensand scarp, has a similar ancient origin, but as yet such

[1] Ekwall, op. cit., under 'Chart'.
[2] *Archaeologia Cantiana*, vol. I, p. 167.

a supposition can only be based on dangerously slender place-name evidence.[1]

The much larger village of Sutton Valence appears to be an exception to the generally dispersed settlement along this zone, but at this point the East–West road is intersected by a fairly important road from Maidstone in the North which proceeds Southwards across the Weald to Hastings and the South coast, and this probably explains the rather greater importance of the village, though even here it will be noticed that the settlement itself is elongated in form, being aligned along the scarp road in an East–West direction.

Although the typical Kentish landscape of orchards and hop-fields can be seen on the photograph, it should be noted that a good deal of this Greensand belt is still being used for arable agriculture.

Evidence on Air Photographs of Former Settlement Patterns

It is not proposed to enter into a consideration of the importance of air photographs from a purely archaeological point of view since this subject is already largely documented and a great deal more work on it is in progress. From a geographical point of view, however, the changed patterns of settlement and communications which are evidenced on the air photographs are of extreme importance and, accordingly, four plates have been included as examples of old settlement sites, ancient routeways and former cultivation patterns, in addition to evidence of former conditions which have already been mentioned when considering the foregoing plates primarily from other points of view.

Perhaps the most striking feature of photograph A on Plate 75 is its remarkably clear demonstration of the nucleated character of the settlement in the village of Queen Charlton which lies about 5 miles to the South-east of Bristol. This impression is greatly aided, moreover, by the way in which one's eyes are focused upon the village by the nearly radial arrangement of the field boundaries round it, which is suggestive of an early agricultural pattern based on the village itself. Closer examination, however, reveals that the modern walls and hedges only form a limited part of the radial pattern and, particularly between the village and point C, traces of old divisions within the modern fields complete the pattern. On the opposite side of the village towards D it seems possible that the narrow modern fields may coincide with the older ones, but in the slightly larger field (E) two ancient divisions again appear, whilst at F, at any rate on the original print, a further group of field boundaries can just be discerned. Thus, detailed examination of the air

[1] *Archaeologia Cantiana*, vol. 45, p. 79.

photograph can be used to build up a reasonably accurate picture of the early field system associated with this old village and the extension of the study to the surrounding country-side may throw light on the true character of the former agricultural settlement of this area.

Photograph B, by way of contrast, shows an area where there has been no continuity of settlement comparable with that shown on photograph A and where evidence of ancient occupation merely serves to emphasize the changed geographical significance of the site. The area lies to the North of Barsolis Hill near Castle Douglas in South Scotland and is a region of hummocky Boulder Clay interspersed with rock outcrops which can never have been of great agricultural value and today is a district of sparse and isolated settlements. During the Anglo-Norman occupation of South Scotland, however, the Galloway region was only held with difficulty, and the occupying peoples found it necessary to disperse small strong points or mottes throughout the region for defence purposes,[1] the remains of which serve as reminders of the former strategic significance of this comparatively isolated part of the country. The example of this type of small defended site shown on Plate 75 is of special interest since it shows an excellent adaptation of the environment, the artificial hill and earthworks being situated on the crest of the largest of a series of mounds of glacial material which form the only well-marked relief features in this area.

Plate 76 shows a section of the Icknield 'zone' in front of the Chalk scarp near Princes Risborough as an example of former lines of communication. Archaeological evidence of sites connected with the Icknield Way suggests that it is of Neolithic origin, but a long period of evolution led to its taking the form of a zone of movement rather than a single trackway, and in many areas its remains consist of two parallel roads known as the Upper and the Lower Icknield Way. It has been suggested that the multiplication of trackways along the Icknield zone may have arisen from the use of paths at different levels during different seasons, or from the use of alternative routes in areas where the line of movement was obstructed, so that in some areas there are a great many more than the normal two tracks. In the area shown on Plate 76 the Upper Icknield Way (marked AAA) consists of alternating lengths of country lane and ill-defined track following the contours of the Chalk scarp face itself, well above the spring line and also well above the line of villages along the foot of the scarp. On the other hand the Lower Icknield Way (BBB) is sited well out on to the plain, below the spring line, and lies along the modern road from Chinnor (C) towards the East. Beyond E, however,

[1] Andrew Lang, *History of Scotland*, vol. 1, pp. 65, 66.

where the Lower Icknield Way crosses the lower edge of the print, it becomes a lane and continues Westwards in a straight line which does not pass through any sizeable settlement. In this area, to the West, the villages of Oakley, Crowell, Kingston Blount and Lewknor all lie above the Lower Icknield Way but are themselves along the line of a slightly higher road appearing on the present print as FF and producing the remarkable rectangular pattern of Chinnor which compares with a similar layout at Kingston Blount and Watlington, near by. It is tempting to suppose that this middle road may lie along the line of an inter-mediate level of the Icknield Way and that the road through Bledloe (D) commencing at G is a possible Eastward continuation of it.

The stretch of Roman road selected for illustration in Plate 77 is a short portion of the Fosseway on the dip slope of the Cotswolds about 10 miles to the North-east of Bath in a direct line between that city and Ciren-cester. From the bottom of the print to A the road has been almost completely abandoned as a routeway and consists of little more than a double line of hedges separated by a strip of grass along which there runs a somewhat ill-defined footpath. For about a mile beyond A the old route is followed by a short section of a secondary road connecting the Cotswold villages of Alderton and Yatton Keynell, but thereafter it again reverts to the condition of little more than a trackway. The virtual abandonment of the Fosseway in this region is an interesting illustration of changing values. Designed by the Romans as a cross-country defence line and link with Eastern England it was planned in this section to traverse the dip slope of the Cotswolds between the Roman centres of Bath and Cirencester without crossing any of the major valleys of the dip slope or scarp-face streams and as such had a remarkably level and straight course. In following this course, however, the Fosseway did not pass through a single future settlement centre between Bath and Ciren-cester and with the end of the strategy for which it was designed its utility diminished and it has been replaced by the modern road which follows the pre-Roman Jurassic trackway (see Plate 73) near the Cotswold scarp, and is well placed to serve the villages along the crest of the scarp and on the Marlstone ledge of the scarp face in the direction of Tetbury and Cirencester, whilst giving access, in addition, to the extremely important settlement centre round the Stroud valleys.

The study of former cultivation patterns is an aspect of archaeological studies in which the use of air photography has already been widely developed, partly because some of the traces of such cultivation marks can only be discerned by these means, and one of the earliest general descrip-tions of ancient cultivation in Britain is well illustrated by vertical air photographs.[1] It is clearly impossible here to attempt the inclusion of all types of evidence of ancient cultivation seen on air photographs, nor is it proposed to enter into the question of the presumed age of the various forms of lynchets and old fields. Plate 78 does, however, show a fairly wide range of cultivation marks. Photograph *A* shows part of the village of Cerne Abbas in the valley of the River Cerne, where a number of different types of cultivation pattern occur within a very small area. At *C*, lynchets of the type already seen on the Chalk and Jurassic scarps mark the end of one of the spurs of Chalk overlooking the valley, whereas round *D* the long, narrow fields, in which even smaller subdivisions can just be identified, mark the development of an early strip pattern possibly based on the acre field. This latter division into acre strips is a form of cultivation pattern which is often recognizable on air photographs and in the arable field near *E* it is just possible to see traces of strips of this type.

Photograph *B* shows the valley known as Ranscombe Bottom in the Chalk of the Marlborough Downs about 10 miles West of Marlborough, and is principally of interest because of the very great lynchet development on practically all the slopes of this group of valleys indicating a former intensive agricultural use of this area, probably in the prehistoric period. In addition to the lynchets following the valley slopes, however, there is a good example of strip cultivation patterns across one of the spurs at *F*.

Road patterns and the actual roadway shapes, together with the often related shapes of field boundaries, form a most important aspect of the cultural landscape and one where the use of air photographs can be of considerable value to the geographer. Examples have already been discussed of road patterns and field boundaries related to physical features, as, for example, the straight lines of recently reclaimed lowlands and the gently curving ridgeways, whilst in addition it will be recalled that former lines of communication, such as the Roman Ermine Street, may profoundly influence these patterns. Nevertheless, even if the reason for the patterns is not a single cause but springs from a variety of factors in the physical geography and history of the area, the result is often a landscape feature which forms one of the most distinctive characteristics of a particular locality. Plates 79 and 80 have therefore been chosen to illustrate two areas of rich agricultural land, without marked relief features, where a different historical evolution has led to totally different cultural landscapes, both in land use and in road patterns and field boundaries. Plate 79 shows an area about 4 miles West of Kelso,

[1] E. C. Curwen, 'Prehistoric agriculture in Britain', in *Antiquity*, vol. 1, pp. 261–89.

immediately to the North of the Tweed, in a region of rich arable land. In this region large settlements occur almost solely in the form of towns at strategic sites on the rivers, and elsewhere the settlement has a pattern of widely dispersed and isolated hamlets and individual farms with a total absence of true village development. Here, comparatively late enclosure into the large arable fields which appear on the photograph has produced a strikingly rectangular pattern, not only of the field boundaries but also of the associated road system linking the farms with one another and with the main roads of the area.

The contrasting landscape of the area of Somerset shown on Plate 80 is immediately apparent on the photograph. The area lies round the village of Winford (A) some 5 miles South-west of Bristol, and has been in agricultural use since very early times, a fact which might be deduced from the traces of old cultivation patterns at C, the signs of acre fields at B and the large numbers of narrow fields elsewhere, probably derived from ancient strips. Here, enclosure for pasture occurred very early in the agricultural revolution and the resultant patchwork of irregularly shaped small fields has been a feature of the Somerset landscape on which comment has frequently been made. In consequence of the early establishment of field boundaries the road pattern was fixed along the lines of the pre-existing lanes among the irregularly shaped fields and is therefore itself characteristically irregular in its detailed shapes and directions, though the marked convergence of these roads on the village of Winford serves to draw attention to the importance of such nucleated settlements in the evolution of the landscape of the English plain.

Settlement Shapes

The fact that an air photograph automatically shows practically all the topographical details of a settlement, including the type and site of individual buildings, field boundaries, vegetation and even such temporary phenomena as current land use and field paths, makes it the ideal medium for the study of settlement shapes. Reference should be made to Plates 14, 51, 67, 68, 69, 70 and 75a, where the photographs show the relationship which exists between special local physical conditions and the distinctive shapes of a number of towns and villages, whilst in Plates 81, 82 and 83 several villages have been included to represent settlement shapes which tend to recur in certain areas, though the influence of the physical environment in these cases is perhaps less direct and obvious.

Plate 81 shows the two Cotswold villages of Tormarton and Yatton Keynell which share the nucleated character of settlement throughout the greater part of the South Cotswold dip slope. Tormarton (A) lies just off the Bath–Cirencester road, a few miles from the Cotswold scarp, and is situated at the junction of the Great Oolite and the Fuller's Earth where a narrow tongue of the latter formation is exposed along a very small valley cut into the edge of the Great Oolite. It has already been pointed out that water supply is limited on the crest of the Cotswolds, and one of the factors leading to the development of a nucleus of population in Tormarton was undoubtedly the presence of springs thrown out by the Fuller's Earth in this locality and the availability of water in comparatively shallow wells, a source still used by some of the farms away from the modern piped supply. It will be noticed that the greater part of the buildings in Tormarton consist of farm units, including barns and outhouses as well as the farmhouses and a small number of cottages, so that the village does in fact represent a true nucleation of the agricultural settlement of this region. There is clearly no well-marked pattern in the settlement, nor is there any obvious convergence of roads on Tormarton, which appears to have grown up along a group of the lanes which divide the large arable fields of this locality.

Yatton Keynell (B) is situated much farther down the dip slope, about 4 miles to the North-west of Chippenham, and lies on a narrow belt of Forest Marble between the main outcrop of the Great Oolite to the West and the Cornbrash to the East. Even here water supply presents a major problem, and until the construction of a piped system the area depended on somewhat unsatisfactory wells near the edge of the Cornbrash, the presence of which may have been a contributory factor in the development of the village. Here, too, the village is obviously a collection of agricultural settlements, though its shape contrasts with that of Tormarton since it has been associated with the intersection of two roads producing a narrow X shape which is frequently repeated in comparable nucleated agricultural villages.

In contrast with the relatively compact villages included in Plate 81, the two photographs on Plate 82 show settlements which are distinctly elongated in form. Although it is true that there is a nucleus of settlement at Langtoft (A) its shape makes it clear that the development of this village was influenced by factors other than a simple congregation at a favourable focal point. The village lies near the edge of the Fens, just above the Car Dyke, about 9 miles North of Peterborough, where the road Northwards from that town, skirting the edge of the Fens, is crossed by a secondary road. It is along the line of this secondary road that the village of Langtoft has developed, and one of the most remarkable features shown in the photograph is the almost complete absence of settlements away from this alignment. Although this is an entirely agricultural region, the village does not consist of a grouping of farms comparable with the Cotswold villages, but is composed almost entirely of houses and cottages with only one or two larger farm buildings. Nevertheless, the location of the village at the road crossing has meant that communication is easy, not only with the drier land to the North, South and West, but also with the reclaimed land of the Fens to the East, where there is an almost complete absence of dwellings, and it seems reasonable to suppose that the concentration of settlement along the road is indicative of the function of the village as a centre of dwellings dependent on ease of access to the surrounding farm lands.

In the case of Yatton (B) the factors which have led to the development of an elongated settlement are entirely different. Here the village is situated on a narrow belt of well-drained land forming a low ridge between two areas of alluvium which were formerly part of the ill-drained marshy region of the Somerset levels (see Plate 52). It seems probable that the initial settlement grew up on the line of a routeway along this ridge and has subsequently extended laterally over the zone of good land of the ridge so that it no longer has the marked alignment of villages like Langtoft, but has, nevertheless, a long, narrow shape governed by the form of the relief feature to which it owes its origin.

It should be pointed out that, unlike the other villages described, Yatton is no longer solely an agricultural centre. The station in the top left-hand corner of the photograph is at a minor rail junction within 10 miles of Bristol and in addition local industries have developed round the village, so that Yatton has become a small residential centre—a fact which might be deduced from the considerable numbers of larger houses, particularly off the line of the main road.

The earlier character and function of settlements is often an important factor in determining their present shape and lay-out, and almost every ancient town involves a pattern marking the old core of the settlement which can be identified readily on air photographs, as, for example, in the case of High Wycombe. Where the old town was surrounded by a defence wall, which marked the limit of settlement for a considerable period, the distinction between the old inner area and more recent development along the roads leading to the town is a very clear one and the photograph A, Plate 83, is an example of such a development in the town of Launceston. Dunheved was the earlier name of the settlement in this strategic position on the Kersey, a tributary of the Tamar, on the Eastern border of Cornwall, and at Domesday it had a castle and a market. The medieval town was built in the area protected by the castle (B) and was surrounded by a wall which results today in the remarkable compactness of the older part of the town shown in the photograph, contrasting sharply with the recent tendency for building to spread outwards from Launceston in a series of straggling extensions along the roads which converge on the town.

Such a tendency for settlements to spread along the line of existing roads appears to be a process which has gained momentum during the last fifty years, and in its most acute form it led to the phenomenon of ribbon development along the main roads on the outskirts of a great many towns and cities. Photograph B on Plate 83 is not an example of ribbon development in the strict sense of the term since it lies 7 or 8 miles outside the city of Bristol and is a relatively isolated phenomenon in this area, but it is included because it is a particularly good example of modern building along even minor roads as well as along the main roads. At C and D the photograph shows portions of the old villages of Farleigh and West Town on the main road from Bristol to Weston-super-Mare, and there seems little doubt that nearness to Bristol and the existence of the main road and main services along it, combined with lower costs and rates, were the chief causes of the development of a continuous line of residential settlement between the two villages. At E, the village of Backwell lay off the lane which later developed into the main Bristol-Weston road, but nevertheless, each of the narrow lanes linking it with the other villages has now been lined with a row of modern houses whilst a similar growth can be seen at F where a side road leads to Nailsea and Backwell station, affording another easy route from this new residential district to the city of Bristol.

Chapter 7

ECONOMIC CONDITIONS AND THE FUNCTION
OF SETTLEMENTS

It is only possible to include a limited number of photographs to illustrate further the influence of the economic rôle of settlements on their location and character, and the examples have therefore been selected to show small towns and villages where a single economic function makes an appreciation of this influence relatively simple, the first three photographs being devoted to various forms of seaside settlement for purposes of comparison. Plate 84 (A) shows two fishing villages on the North coast of Cornwall, about 6 miles North of Wadebridge. The general decay of the small-boat fishing industry of North Cornwall has led to the complete abandonment of Porth Gaverne (D) as a fishing centre, but a number of boats still operate from Port Isaac (C) which is now one of the few remaining fishing villages of this type in Cornwall. The suitability of the sites of Port Isaac and Porth Gaverne at the top of small shingle beaches in well-sheltered bays calls for no comment, but it should be noticed that the bays form the seaward end of two deep valleys well below the general level of the coastal plateau of North Cornwall and access on the landward side is difficult. It is therefore interesting to notice that whereas Porth Gaverne has virtually disappeared and the original village of Port Isaac, huddled in the valley, remains a model of an unspoilt fishing village, the plateau-like area between the two valleys, on the other hand, has been the scene of a great deal of building during the last fifty years and it is in this area, which is technically part of Port Isaac, that most of the hotels and boarding-houses are situated, forming the basis for the small tourist industry which is now the area's main source of wealth. The factors of access and communications still accentuate the difference between the two parts of Port Isaac, since the upper area (E) has quite easy access to the main roads of North Cornwall and to the railway at Port Isaac Road and has bus services operating from the point F, whereas beyond this point the roads are extremely steep and narrow and the route through the old village towards the top right-hand corner of the photograph is generally regarded as being unsuitable for vehicular traffic.

Photograph B has been chosen to provide the greatest possible contrast with the old settlement of Port Isaac and shows an area immediately to the West of the residential centre and resort of Bognor Regis. Here the development of settlement is not complicated by relief and the good level building land both to the East and West of the town has permitted the planned development of residential areas behind the long, straight beaches of this portion of the South coast. The locality shown on the photograph is known as the Bay Estate and it will be seen that it is highly characteristic of many seaside towns and estates which have been built specifically as resorts or residential areas, since its lay-out has been based on a rectangular road pattern, aligned with the beach, serving a settlement type composed almost entirely of well-spaced detached or semi-detached houses. At G it will be appreciated that the main road of the estate and a number of side roads have been made up beyond the present limit of building construction, an indication of planned future developments, whilst beyond the edge of the print the road G leads to a comparable settlement area at Pagham Beach Estate on the Eastern side of Pagham Harbour.

Plate 85 shows the mouth of the Looe River and the twin towns of East and West Looe which may be regarded as intermediate types of coastal settlement between the extremes represented by Port Isaac and the Bay Estate. The Looe River provides a natural harbour which has long been used by small craft, and before the development of the tourist industry Looe had an appreciable local importance as a small port and market centre. In this case, however, the development of the tourist industry has been appreciably greater than at Port Isaac, partly because of its position on the South coast and partly because of its direct rail link (A) along the river valley to the main line at Liskeard. The small beach (B) and a number of other local bay beaches may partly account for Looe's development as a resort, though in addition the scenic attractions of the river valleys and the use of the river as a base for pleasure craft are probably quite as important. The continued importance of the river valley, whether as a harbour, as the main line of road and rail communications or as a tourist attraction, is borne out by the form of the town itself, there being practically no building development away from the valley-sides, except along the favourably situated tributary valley at C and along the coast at the bottom of the print, and indeed one of the most remarkable features shown on the photograph is the absence of any spread of settlement on to the level plateau areas above the valleys.

The pattern of zigzag lines at D represents the position of defences erected to protect the important look-out and defence position on the hill overlooking the river mouth and is an interesting example of the ease of identifying even the slightest structures of this type on air photographs.

A number of market towns have already been included, but in most cases their character has been modified by other developments and Plate

86 has therefore been included to show Chipping Sodbury, which remains an almost diagrammatic example of a Gloucestershire market town despite the encroachments of other economic activities. The name 'Chipping' implies the ancient function of this settlement as a market centre, and the extremely wide main street in which the markets are held still retains much of its original character. On either side of this street the photograph shows the remarkable concentration of the old buildings achieved by their construction with the long axis of the larger buildings at right-angles to the line of the street, and the compactness of the old town contrasts sharply with the much more scattered modern building between it and the railway line to the South. The basis of Chipping Sodbury's early importance was, of course, almost entirely agricultural, and the influence of the town on land use in this area has remained to the present time in the existence of the Sodbury Commons on the relatively poor lands of the Lower Lias Clays to the North-east of the town, parts of which are to be seen in the top right-hand corner of the photograph.

The most striking economic development in this area in more recent times has been the cutting of the very large quarry immediately to the North of the town where the Carboniferous Limestone, forming the rim of the Gloucestershire coal-field, is being excavated on an enormous scale. The industry is obviously a purely extractive one, however, and very little surface plant is necessary, so that, although the quarry appears on the air photograph to have modified the landscape to a very considerable extent, it is nevertheless possible to pass through Chipping Sodbury on the main road without being aware of its existence. On the other hand, at *A* there is evidence of a still more recent tendency for industries to be developed in this locality, in part because of the availability of a suitable labour force in the period immediately before the war. The factory in question is used for the manufacture of cardboard and containers and represents a spread of the very extensive paper and cardboard industry of Bristol. In the main, however, industrial and commercial employment for the population of Chipping Sodbury is in Bristol or at Yate (about a mile beyond *A*) where sizeable light engineering and electrical engineering works have been developed alongside the main Bristol–Gloucester–Birmingham railway line.

The existence of an old nucleus of population, the possibility of light industrial development and the nearness to Bristol (45 minutes by bus) has led to a certain amount of modern building round the old town, but more particularly between it and Yate (beyond *A*), and has produced repeated suggestions that this area should be deliberately developed as an urban centre.

Plate	Locality	Ordnance Survey Sheet no. 1:63,360	Latitude and Longitude	Sortie number	Print numbers
81A	Tormarton Gloucestershire	156	W 02°20' N 51°30'30"	106G UK 1416	4169
81B	Yatton Keynell Wiltshire	156	W 02°11'30" N 51°29'	106G UK 1416	3155
82A	Langtoft near Peterborough	123	W 00°20' N 52°42'	106G UK 1489	4407
82B	Yatton Somerset	165	W 02°49' N 51°23'	CPE UK 1869	4141
83A	Launceston	186	W 04°21'30" N 50°38'	3G Tud. UK 146 pt I	5071
83B	Backwell near Bristol	156	W 02°44' N 51°25'	CPE UK 1869	3110
84A	Port Isaac Cornwall	185	W 04°50' N 50°35'30"	CPE UK 1999	1012
84B	Bay Estate Bognor Regis	181	W 00°43'30" N 50°46'30"	541/217	4026
85	Looe	186	W 04°27'30" N 50°21'	CPE UK 1794	3029
86	Chipping Sodbury Gloucestershire	156	W 02°23'30" N 51°32'	106G UK 1416	3386
87	Bugle Cornwall	185	W 04°47'30" N 50°23'30"	CPE UK 1999	4136
88	Creetown South Scotland	Scotland 91	W 04°22'30" N 54°53'	106G Scot UK 42	4076
89A	Royston Station near Barnsley	102	W 01°26' N 53°36'	541/31	3380
89B	Carberry Tower Midlothian	Scotland 74	W 03°01' N 55°55'	106G Scot UK 119 pt II	5122
90	Treherbert South Wales	154	W 03°32' N 51°40'30"	CPE UK 2081	4246
91	Tipton near Birmingham	130	W 02°02' N 52°32'	541/15	3093
92A	Hawkesbury Upton Gloucestershire	156	W 02°19' N 51°35'	106G UK 1721	3183
92B	Near Chester	109	W 02°44' N 53°07'30"	CPE UK 1935	1191
93A	Near Glengorm Castle, Mull	Scotland 53	W 06°10' N 56°38'30"	CPE Scot UK 275	3052
93B	Harry Stoke near Bristol	156	W 02°33' N 51°30'30"	CPE UK 1869	4017
94A	Cobbins Brook Lea valley	161	E 00°02' N 51°42'	541/183	4038
94B	Five Wents near Maidstone	172	E 00°35' N 51°14'30"	541/536	3211
95	Bath	156	W 02°22' N 51°23'	106G UK 1522 pt II	6119
96	Roch valley near Todmorden	95	W 02°06' N 53°41'30"	541/27	4075

Plate 81

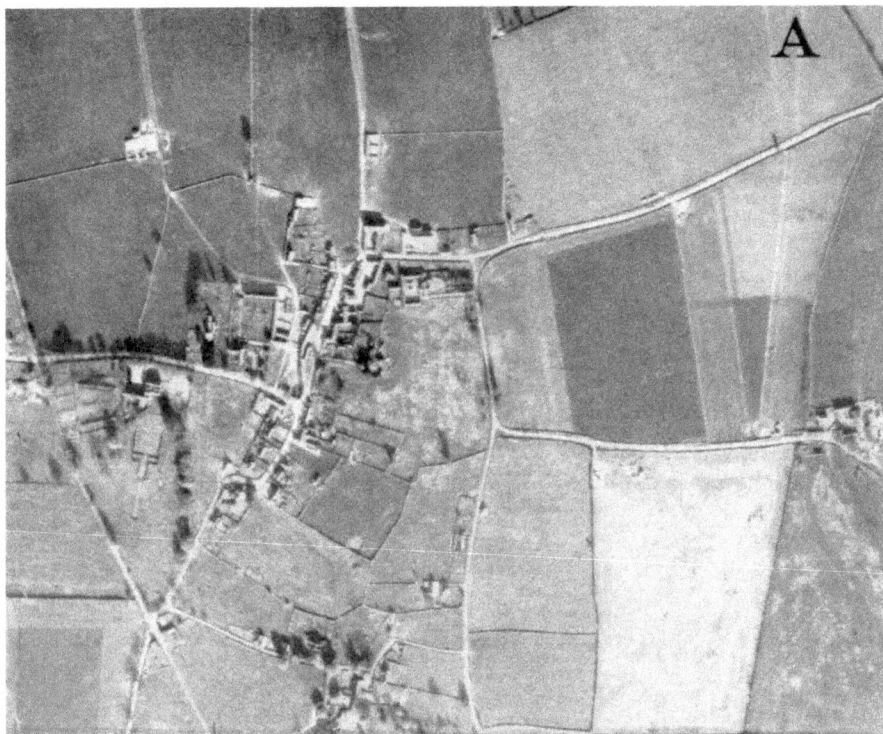

Plate 82

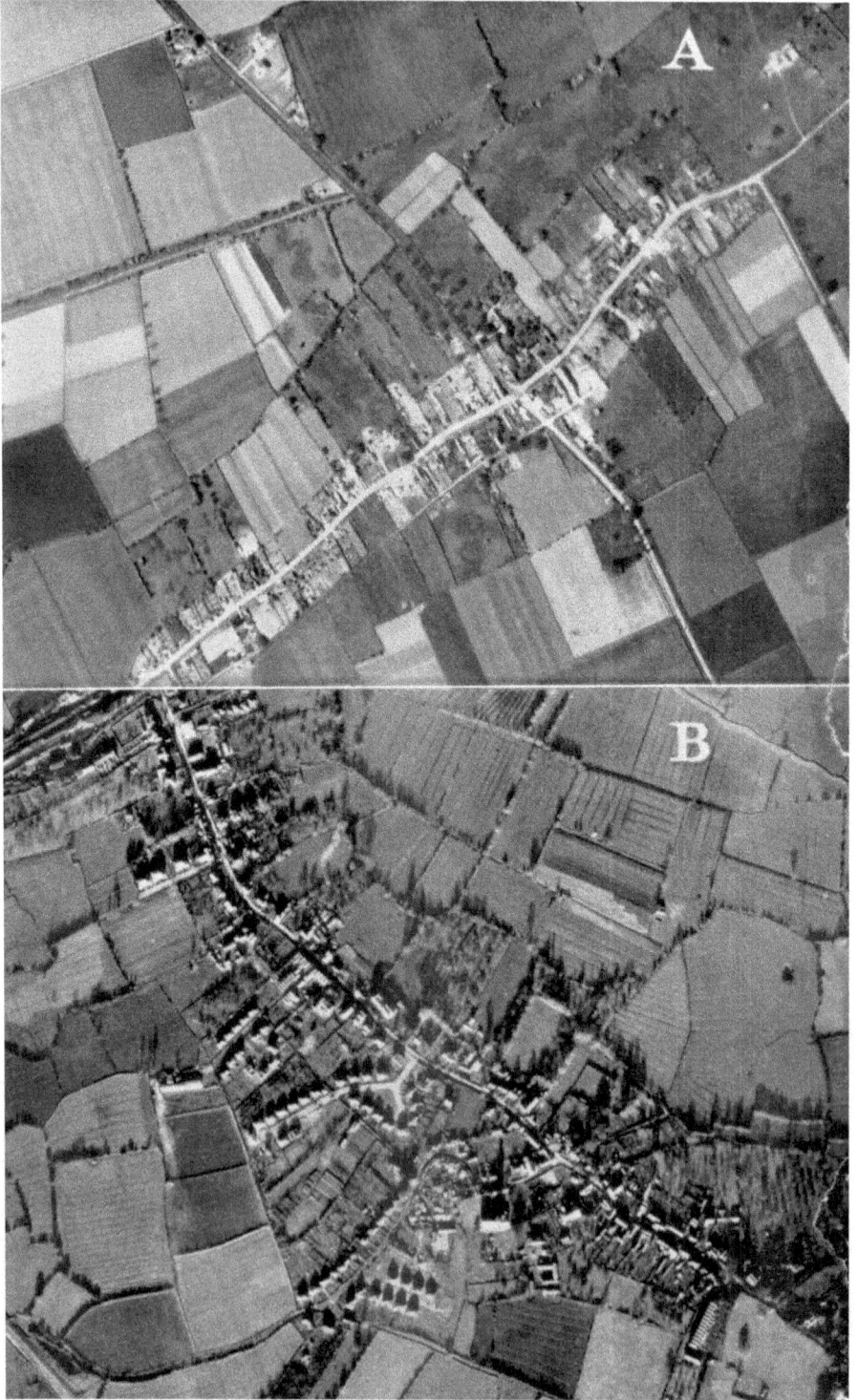

Plate 83

Plate 84

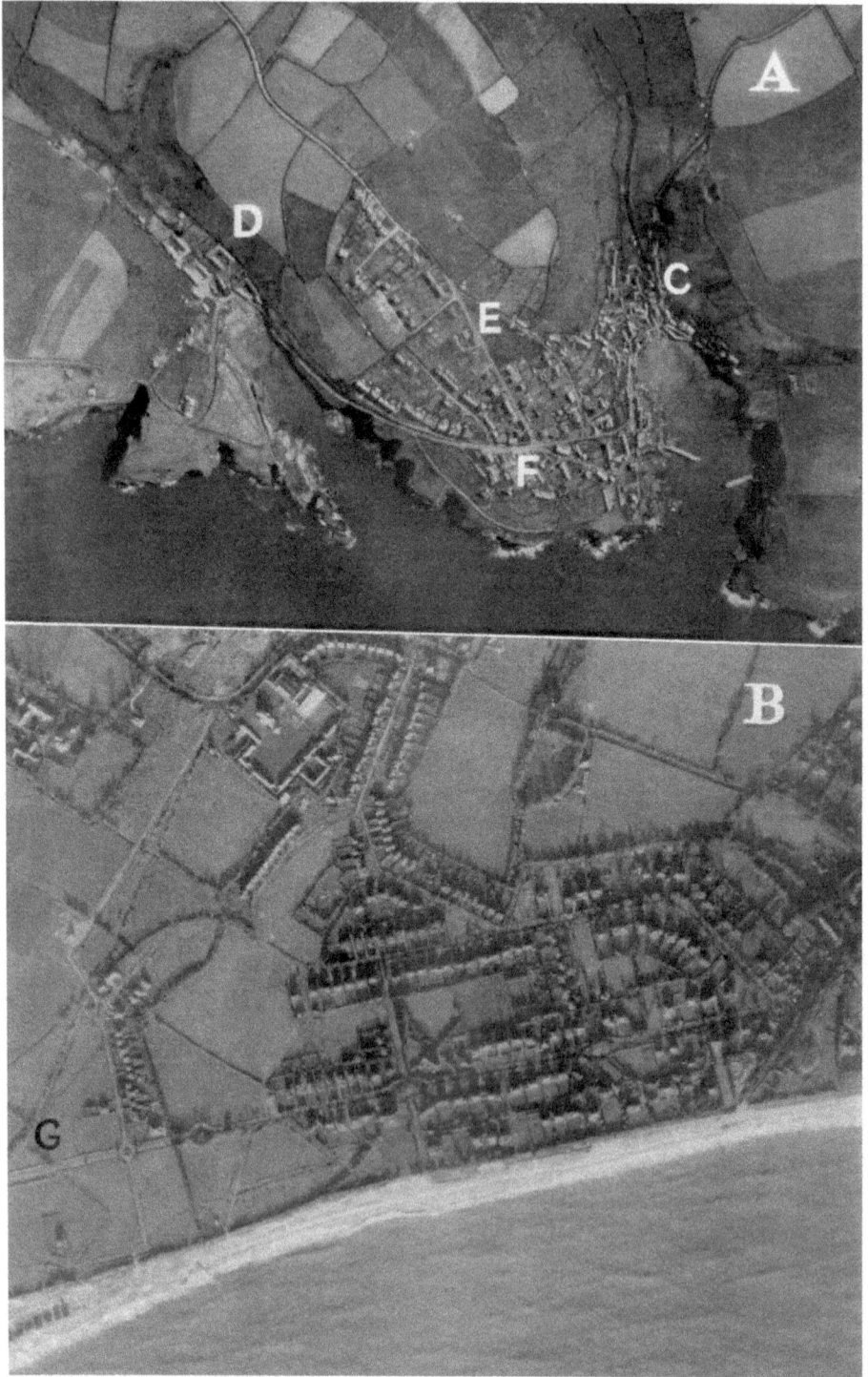

Plate 85

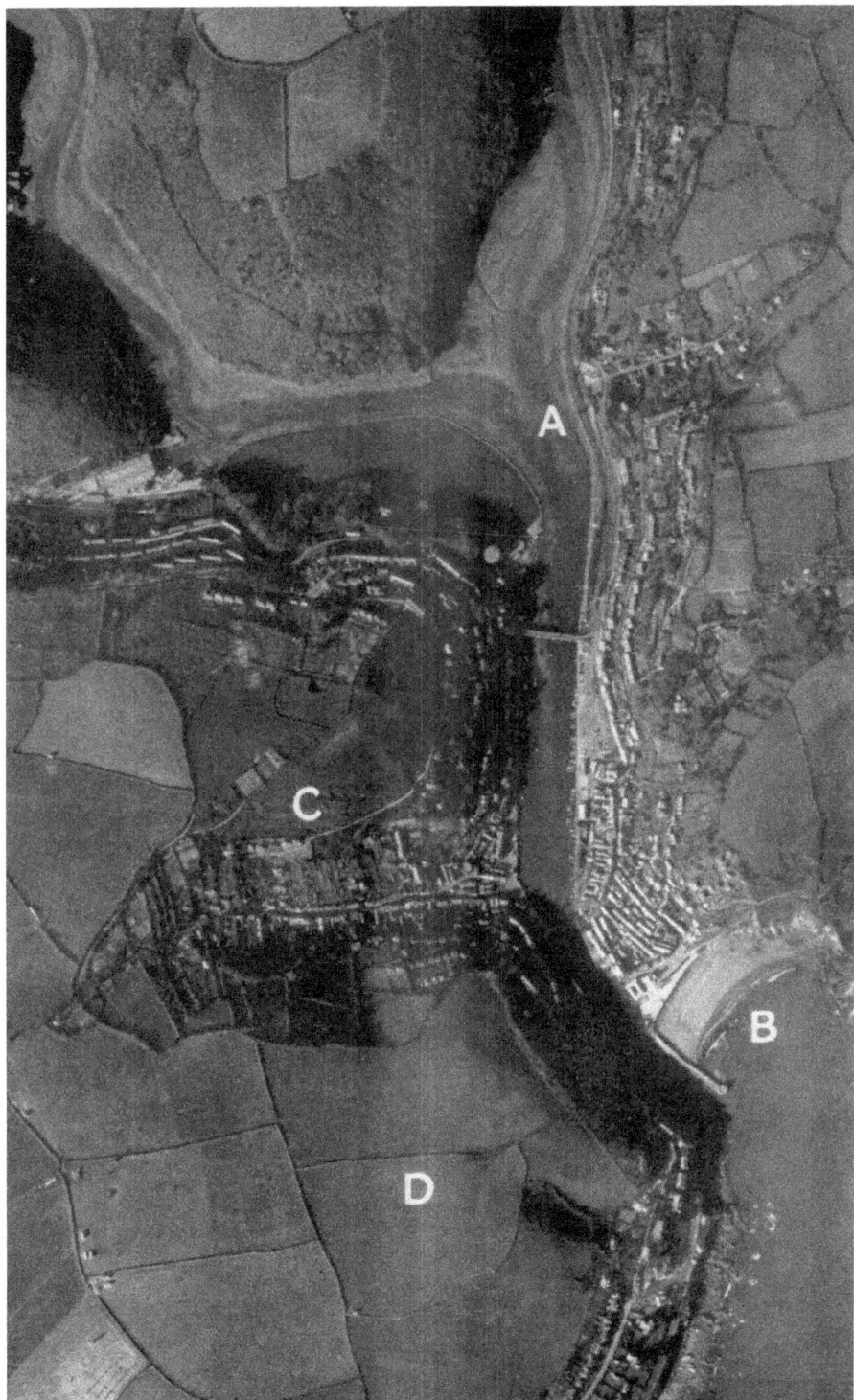

Plate 86

Plate 87

Plate 88

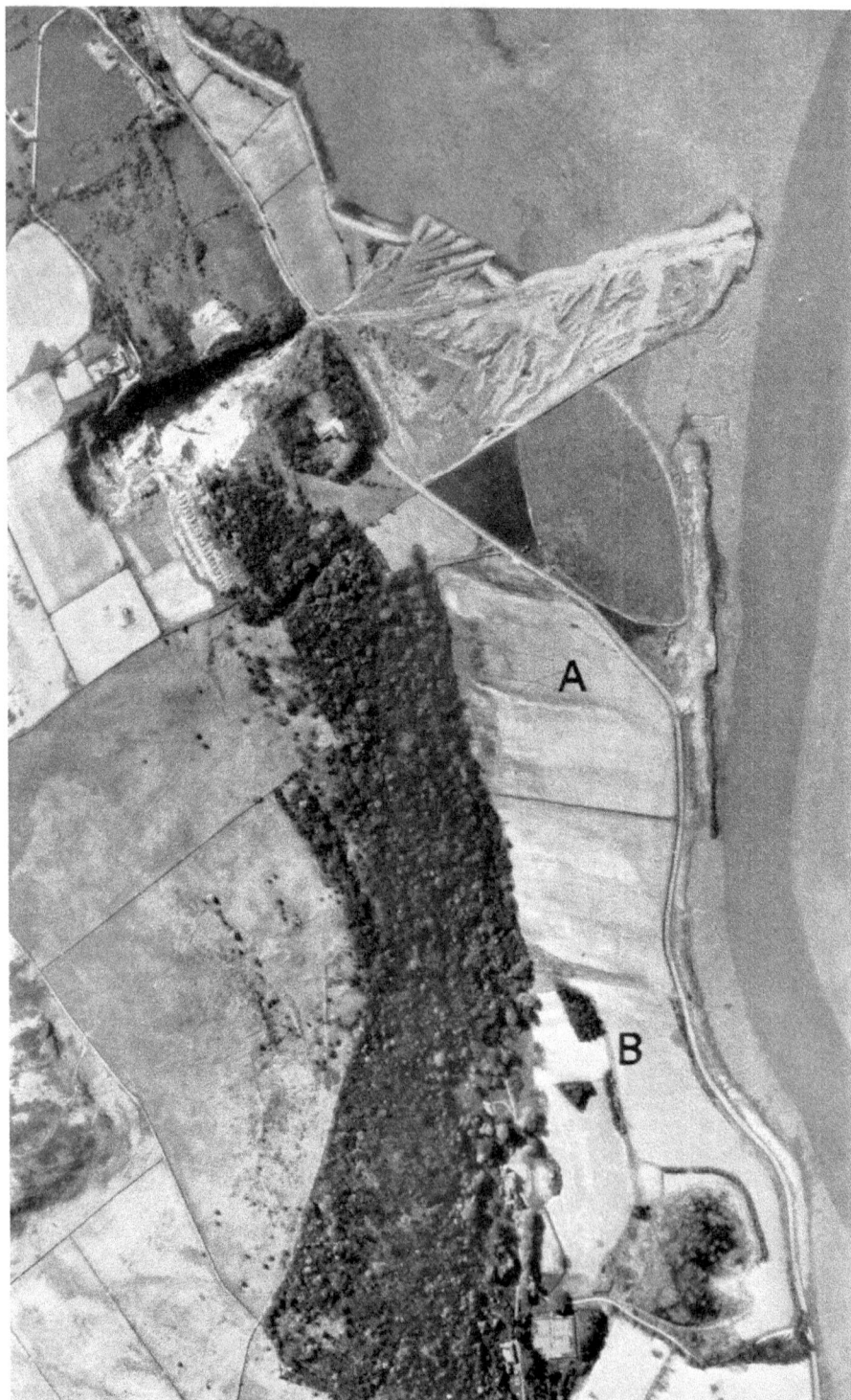

Plate 89

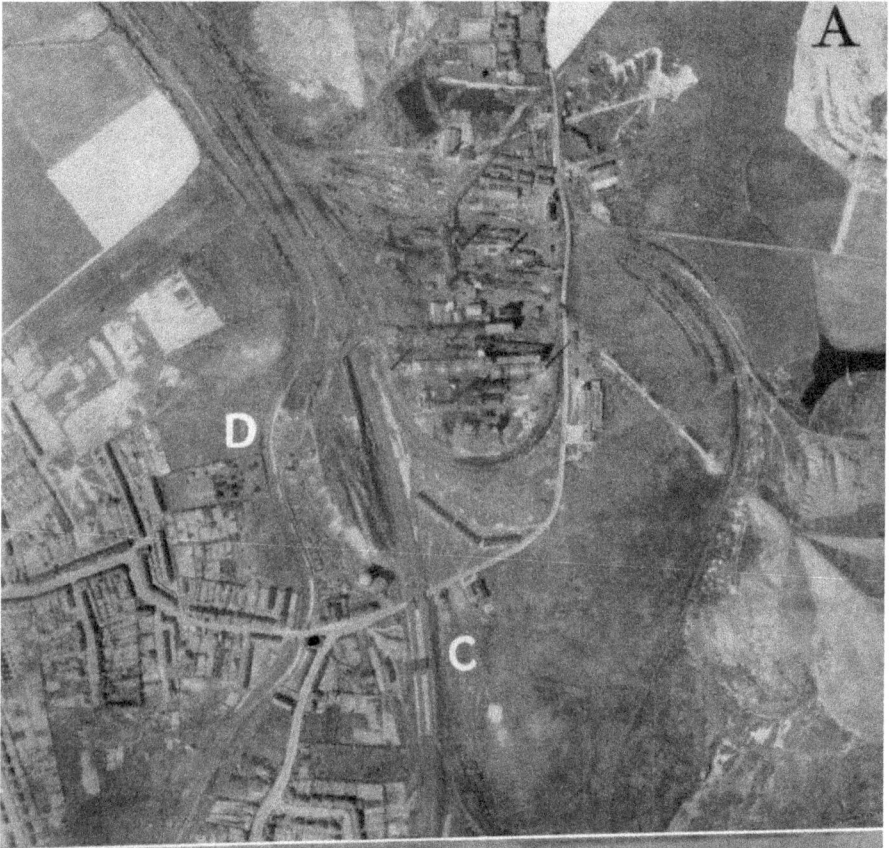

Plate 90

Plate 91

Plate 92

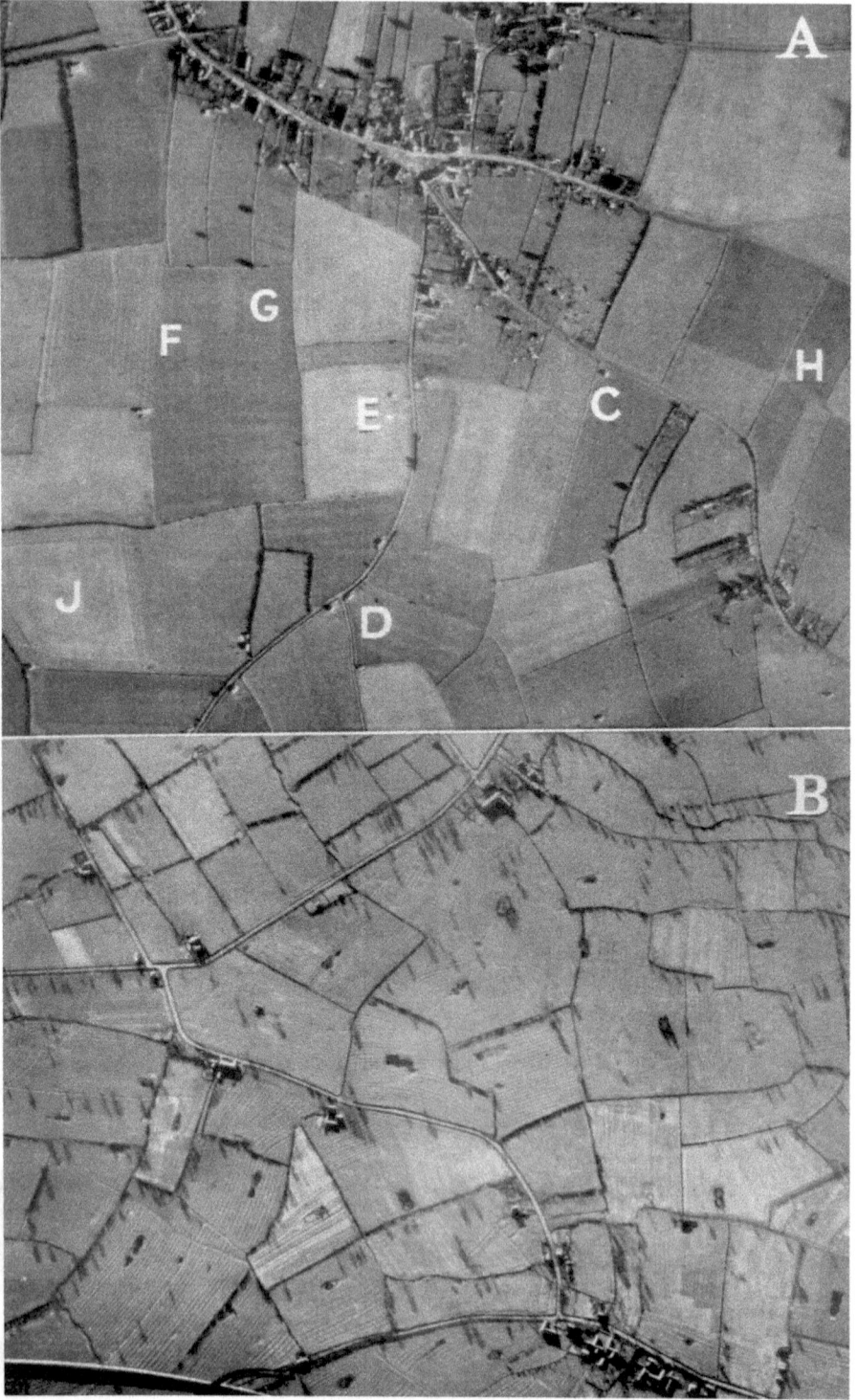

Plate 93

Plate 94

Plate 95

Plate 96

It should perhaps be pointed out that the railway line which appears on the photograph has had comparatively little effect on the recent development of Chipping Sodbury. This is a portion of the direct London–South Wales line via the Badminton tunnel through the Cotswolds and the Severn Tunnel and is used primarily for through goods traffic and express passenger trains. Chipping Sodbury station, which is beyond the limit of the print to the East, handles a certain amount of heavy goods and parcels traffic, though for most local traffic, particularly from the quarry, road haulage is much more important, and the extremely limited passenger train service cannot compete with the very frequent bus service from the centre of the town to Bristol.

Plate 87 is included as the last photograph in this section to show a settlement whose character is entirely dominated by a single economic function. It shows the village of Bugle, on the road from Bodmin to St. Austell, about eight miles South-west of Bodmin in the centre of one of the main Cornish china clay districts. The effect of the china clay working on the landscape of this locality is too obvious on the photograph to call for further comment, but it should be noticed that the lines of communication provided by the Bodmin–St. Austell road and the branch railway leaving the main line in the top right-hand corner of the print have been of great importance to this industry in transferring its product to the coast or to rail connections with the Midlands. It is therefore interesting to notice that, although there is a slight concentration of settlement within the large curve of the railway line, the greater part of the settlement is obviously placed as close as possible to the actual workings and is consequently strung out in an almost continuous line along the main road.

Economic Activity

The study of economic activity as shown on air photographs may take on as many varied aspects as there are different approaches to the study of economic geography, and with a limited number of plates it is only possible to illustrate one or two examples of the main forms of industrial and agricultural activity. Accordingly Plates 88, 89, 90 and 91 have been selected to illustrate a single theme—that is the increasing effect upon the landscape which is produced by the increasing intensity and elaboration of industrial activity. Plate 88 shows the simplest case of a granite quarry to the South of Creetown on Wigtown Bay in South Scotland, and is designed to demonstrate that an extractive industry of more than purely local significance may nevertheless have comparatively little effect on the landscape. The location of the Creetown granite quarry is of some importance since it facilitated the coastwise shipment of large granite blocks, numbers of which were used in the construction of the docks at Liverpool, whilst the shipment of smaller rectangular blocks or 'setts' provided street-paving material for the Lancashire towns during the later part of the nineteenth century. It is interesting to notice that the disposal of 'spoil' from the quarry has been arranged to counteract the silting up of the mouth of the Cree and provide a jetty along the deep-water channel so that coastwise shipments may still be carried on. The buildings on the jetty house crushing and sorting plant for the production of ballast and fine chippings on which the industry has depended in recent times since the virtual abandonment of granite as a building material.

Reference has already been made above to the existence of raised beaches on this portion of the Scottish coast and it may be noticed that on this photograph the junction of successive beach levels can be seen very clearly indeed at *A* and *B*.

Photograph *B* on Plate 89 shows conditions which are in many ways comparable with those at Creetown since it will be seen that the coal-mining industry in this district has produced only limited local effects upon the landscape and land use. The colliery is situated near the Eastern boundary of Midlothian to the South of Musselburgh, and the photograph shows part of the grounds of Carberry Tower at the bottom of the print from which the colliery name of Carberry Mains is derived. Even in the immediate vicinity of the colliery and its spoil-heaps the rural country-side is virtually unaffected by industrial development and demonstrates the characteristics of the Lothian region with its predominance of large arable fields and its considerable numbers of country estates and sizeable farms. Here, of course, the coal industry is simply one of extraction without any development of processing or secondary industrial activity, so that apart from the pit-head buildings and the spoil-heaps the only visible consequence of the opening of the colliery has been the construction of the short length of rail track by which the coal is transferred to the main line, at the bottom left-hand corner of the photograph, for transit to the coast and to Edinburgh.

Photograph *A* on the same plate, however, shows a further stage in the modification of the landscape by the development of the coalmining industry. The colliery is located near to Royston and Notton station (*C*), nearly midway between Wakefield and Barnsley, and the area shown on the photograph is typical of conditions which are developing in this very important part of the South Yorkshire coal-field. Here operations are on a much larger scale and the wide dispersal of spoil-heaps makes them an important feature of the landscape, whilst much of the land not actually in industrial use has become derelict and considerable areas round the industrial plant are out of agricultural use. The importance of this large colliery has led to the development of an elaborate system of railway sidings to serve it, whilst the general importance of the South Yorkshire industrial area and the near-by West Riding in the era of canal construction is indicated by the presence of the Aire and Calder Navigation Canal (D). In this case it will be seen that the colliery buildings themselves are more elaborate and extensive than in photograph B, in part because of the presence of coal processing and carbonization plant associated with the use of the coal in near-by industrial areas, a feature common to many of the larger collieries of the Yorkshire coal-field. Moreover, the considerable industrial developments of this region have had an important effect on its settlement pattern, and in the bottom left-hand corner of the photograph a portion of a residential area appears where there are a number of the long terraces of houses common throughout the industrial towns and mining villages of South Yorkshire.

Plate 90 represents the extreme case where activity based on the development of a single type of industrial function has come to dominate both the landscape and the economic life of the area completely. The photograph shows the mining centre of Treherbert in the Rhondda valley, about 10 miles to the North-west of Pontypridd, and is entirely typical of a very large number of such settlements throughout the valleys of the South Wales coal-field. In this case the features shown on the photograph are occasioned as much by the relief of the area as by the form of economic activity which prevails. The deep and very steep-sided valley provides only limited amounts of land suitable for building, and in consequence the settlement has developed an elongated pattern along the valley floor and at the foot of the steeper slopes of the valley sides. Apart from early surface workings the main collieries are situated near the valley bottom itself, at A, B and C, for example, around which the valley sides, and in one case the floor of the valley (at D), have been covered with large spoilheaps which greatly restrict the available building land and, in more recent times there has been an increasing tendency to carry the spoil by ropeway to the uplands above the valleys. The railway line through the valley and the necessary colliery sidings take up an appreciable part of the lowest ground, and in consequence the settlement of Treherbert consists of little more than three or four near-parallel lines of terrace houses on the lower slopes above the main road along the valley. The fact that, at E, one branch of the road leaves the line of the valley is an indication of the nearness of the head of the Rhondda valley, and beyond the limit of the photograph the road continues to climb up the valley side to cross the moorlands on the rim of the coal-field and join the main roads in the valleys to the North.

The modification of the landscape by the progress of economic development may perhaps be said to culminate in Britain in the West Midland region known as the Black Country where all other features are masked by the effects of industrialization, though somewhat similar conditions occur in relatively restricted localities within the industrial areas of Lancashire, Yorkshire and the lower Clyde. Plate 91 shows a small area two miles to the East of Tipton between Dudley and Wednesbury which seems reasonably typical of the region as a whole. The first impression gained from the photograph is that of a confused and shapeless distribution of industrial plants and small dwellings and an irregular pattern of roads, interspersed with considerable areas of apparently derelict land partly occupied by surface excavations many of which have been abandoned and partially flooded—an impression which coincides very closely with that obtained when passing through the area by rail. When it is remembered, however, that much of the industrial development of this region belongs to the period of canal and railway construction it will be possible, by closer examination of the photograph, to appreciate that a certain pattern does exist which again is reasonably characteristic of comparable districts elsewhere. This pattern consists of two branching systems of canal and rail transport lines forming part of the elaborate and dense network of canals and railways between Birmingham and Wolverhampton, and it will be found that every major industrial unit which appears on the photograph lies alongside one of these lines of communications. Roads were of comparatively slight importance in the industrial evolution of this area and the wide main roads which now traverse the area have been recently reconstructed, usually along the line of paths and tracks which linked the factories with the older groups of dwellings at A, for example. The important road system of the area, which now provides the industrial plants with road transport facilities, appears, therefore, to have been superimposed upon the older patterns of rail and canal transport, and this double pattern is particularly clear in the top left-hand corner of the print where road reconstruction is linked with the creation of new by-roads for an area of recent house-building.

The use of air photographs in the study of agriculture and in agricultural research offers considerable scope for detailed application of the techniques of interpretation. It has already been shown that variations in soil character may be appreciated and their effects on vegetation and agriculture assessed, whilst in addition the effects of aspect, slope and exposure have been mentioned. With the aid of farm plans it is possible to study the details of farming technique, the variation of crop quality within individual fields and, where repeated photography is possible, a comprehensive picture of land use may be built up. From a more general geographical point of view, however, air photographs may also be used to illustrate the landscape associated with the major types of agricultural practice, and in the present section the succeeding photographs are designed to show six areas where comparatively simple and uniform land utilization can be illustrated.

Photograph *A* on Plate 92 shows a small area on the Great Oolite of the Cotswolds at the village of Hawkesbury Upton where the land is entirely in arable use. The photograph was taken in the autumn of 1946 when a great deal of the land in this area was being harvested from grain crops. At the edges of the fields (at *C, D* and *E* for example) stacks of hay from the ley fields may be discerned, but the cut grain is, of course, normally taken direct to the stack-yard. At *F* and *G* two fields have been cut and the stooks are still standing, whilst at *H* some of them have already been carried. At *J*, on the other hand, a field is in the process of being cut and several rows of stooks are visible round the outer edge of the standing crop. The village of Hawkesbury Upton is typical of the compact villages of the crest of the Cotswolds (cf. Tormarton, Plate 81(*a*)) and its origin may be related, in part, to the presence of wells which yield a reasonably good water supply in this vicinity.

In complete contrast, photograph *B* shows a portion of the Cheshire plain some eight miles South-east of Chester near the Chester–Crewe railway line which appears in the bottom left-hand corner of the print. Here the smaller fields with their hedges and trees are responsible for some of the contrast with the Cotswold area of large fields divided by low stone walls. In addition, however, it should be noticed that there are only minor variations in colour between the fields in the Cheshire area which is entirely given over to grass farming for the maintenance of dairy herds. The very long shadows on this particular photograph give prominence to the furrow marks which are indicative of the former use of a great many of the fields, whilst they also show up very clearly the marl pits near the centre of almost all of the fields. These shallow depressions, often partially filled with water, are a very characteristic feature of the Cheshire landscape and are a relic of the former practice of marling the land with material excavated at the most convenient point for distribution over the whole field. There is a small grouping of buildings in the hamlet of Newton at the bottom of the photograph, but in general (apart from small market towns) the dispersal of fair-sized farms and occasional cottages along the roads of the area is typical of the settlement pattern throughout much of the Cheshire plain and is in marked contrast with the concentration of settlement in nucleated villages on the Cotswolds.

Photograph *A* on Plate 93 shows a small area of cultivation behind Port Chill Bhraonain, a little to the West of Ardmore Bay in Northern Mull, which contrasts sharply with the good farming land already illustrated and is typical of the agricultural pattern of crofting settlements in the West of Scotland and in the Isles. The fields consist of no more than irregularly shaped areas of poor soil among the rock outcrops, and at the time of photography a crop of hay had recently been gathered and was still standing in small stacks scattered over the fields. The presence of a good road at the right-hand side of the photograph and the existence of planted woodland is to be related to the location of Glengorm Castle, just beyond the bottom edge of the print, and not to the agricultural character of the area.

Photograph B, at the other extreme, is designed to demonstrate some of the agricultural consequences of a location near the outskirts of a large town or city. It shows the hamlet of Harry Stoke which lies about four miles to the North-east of the centre of Bristol and is only a short distance beyond the limit of continuous building in the suburban area. In common with most large urban areas, Bristol is surrounded by a belt of country where the basic form of land use is grass farming associated with the provision of a milk supply for the town, and most of the area shown on the photograph conforms to this pattern. In addition, however, at C and D there are two fairly large developments of market-gardening which supply vegetables and fruit to the Bristol market, and again the examples shown on the photograph are part of a nearly continuous belt of such farms and smallholdings which surround the city.

Photograph *A* on Plate 94 also shows the effects of nearness to a large urban market; in this case, London. The area shown is on the Cobbins Brook, a tributary of the Lea near to Waltham Abbey, and is again primarily a region of permanent grass farming, though in this case the additional agricultural activity fostered by nearness to the London market is the highly specialized one of cash-crop production under glass, a form of culture particularly associated with the Lea valley.

Photograph *B*, again comparatively near London and to some extent still influenced by it, is mainly intended to show a landscape which is almost entirely dominated by concentration upon horticulture and in particular upon fruit production. It covers an area near Five Wents to the South-east of Maidstone and shows the highly distinctive appearance of the fruit- and hop-growing districts of Kent.

The study of communications and transport systems forms an integral part in the investigation of the human geography of an area and of its relationship with the physical environment and consequently such features have already been discussed in connection with earlier photographs wherever the communication system of the area has had features of general geographical interest. The two remaining photographs devoted specifically to this subject have therefore been selected to illustrate two main functions of air photographs in the study of transport, firstly the demonstration on large-scale photographs of the actual design and lay-out of transport facilities, and secondly the illustration of the use of a natural route by various types of transportation.

For the first purpose a photograph at the very large scale of approximately 1 : 4,600 has been used to show transport arrangements near one of the passenger railway stations in Bath.

From the point of view of water-borne transport the photograph is unsatisfactory, since the river at Bath is seldom used except by very small pleasure craft and barges, but security precautions make it impossible to include large-scale photographs of important ports, docks or canals in Britain. Nevertheless, at C, it is possible to see details of structures on the river front, of small landing places and of the railway bridge and its central pier and it will be apparent that similar photographs of an important harbour or dock would show even the smallest object on the water-front, the arrangement of dock and lock gates and bridges and would afford precise information about the type and size of vessels actually using the port at the time of photography.

The photograph is more satisfactory in showing rail transport arrangements though similar reasons prevent the inclusion of large-scale photographs of major rail junctions, marshalling yards or large city termini. The passenger station at A is a relatively small one, but the rail network associated with it is sufficiently elaborate to demonstrate the ease with which the lay-out of the tracks can be traced on photographs of this type. It will be seen that the track cross-overs at the approaches to the station are visible, and near the bridge over the river even the individual sleepers can be detected clearly under a hand lens. Alongside the station a number of passenger coaches are standing on a short siding outside the main station buildings. The individual tracks and systems of points are clearly visible in the small marshalling yard on the opposite side of the river near the engine shed B, and a single track can also be traced passing along through the buildings to G where a number of trucks are standing within the industrial plant at the side of the main line. Elsewhere in this area numbers of wagons are to be seen, and in most cases it is possible to identify the various types of open and closed vehicles. The circular object near the engine shed is a locomotive turntable.

It is obvious that the general pattern of roads can be traced on air photographs at almost any scale, but in this case it is possible, in addition, to study the actual traffic arrangements within this part of the city. For example, the traffic marks on the roads are everywhere visible, and at D there is a remarkably clear case where an island at a road fork results in the traffic lanes forming a network of tracks round the white mark of the island itself. Round the rectangular block of buildings between E and F a "one-way" system of traffic control is in operation, and near F it is possible to see the arrows painted on the road to indicate the compulsory line of movement, which is being followed by a small open truck. At the other end of the "one-way" circulation, at E, it is possible to read the lettering "No Entry" painted on the road and near to this a "No Parking" sign on the roadside is equally visible, as are the white lines at "Halt" signs and at traffic lights throughout the whole of the area on the photograph.

Plate 96 shows the system of road, rail and canal transport in the Pennine gap formed by the valleys of the Roch and the Calder which acts as one of the main routes linking South-east Lancashire with the West Riding of Yorkshire. The photograph covers a part of the Roch valley, on the West side of the Pennines, immediately downstream from Todmorden where the river turns Southwards. The canal is the Rochdale canal linking the water-borne transport system of South-east Lancashire with Todmorden and Halifax, whilst the railway is the main line of the former Lancashire and Yorkshire railway from Liverpool and Manchester to the West Riding and Leeds. It will be noticed that the canal has no less than ten locks in the area shown on the photograph, raising the water level towards the crest beyond Todmorden, whilst the railway at A passes through one of a number of short tunnels which enable this section of the line to cross the Pennines through the Roch and Calder valleys without undue curves or gradients and with a minimum of tunnelling.

The importance of the road, rail and canal systems in the narrow valley of the Roch is emphasized on the photograph by the remarkable concentration and continuity of industrial development along the valley. Starting initially as stream side sites during the early stages in the evolution of the textile industries in the West Riding and South-east Lancashire, the continued existence of this thin ribbon of industrialization has been largely ensured by the excellent system of communications which passes along it, linking the major industrial areas to the East and West of the Pennines.

APPENDIX

Photogrammetry

In order to use air photographs for survey purposes or for the precise measurement of heights and distances it is necessary to devise means of solving three problems:

1. To orient successive photographs correctly in relationship to one another. This can only be done when a solution is found to problem 2.
2. To eliminate, or allow for, errors arising from the fact that the optical axis of the camera is never, in practice, truly vertical to the plane on which the survey is to be carried out (in other words the camera is tilted) nor is the flying height absolutely constant.
3. To make allowance for variations in ground height, that is the varying distance between the camera lens and the topographical detail to be mapped.

Without the use of moderately elaborate equipment it is only possible to arrive at an approximate solution to these problems. If there are only very slight tilts on the air photographs and the maximum variations of ground height are only a small fraction of the flying height, it is possible to use semi-graphical methods of mapping from air photographs which do not solve the above problems completely but, nevertheless, give satisfactory results, provided that a fairly dense network of ground control is available, but these methods are not suitable for precise work. However, when the photographs have been oriented by such a method it is possible to make height estimations, using the parallax bar (photograph A, Plate 7), which would be suitable for the measurement of larger relief features. Variations in height are represented on the oriented prints by variations in the distances between the corresponding images on the two photographs and these differences are recorded on the micrometer screw (A) as the engraved marks in the centre of the glass plates BB_1 are moved to lie over the two images of the objects in question. Absolute heights may, of course, be obtained by reference to other spots of known height which act as control points.

Photograph B on Plate 7 and Plate 8 are two examples of equipment designed to solve simultaneously and completely the three problems mentioned above. The Wild Stereoautograph shown on Plate 7 is shown as a stereoscopic pair of photographs so that the machine may be examined under a hand stereoscope. The two 'cameras' A and B are so arranged that the overlapping photographs mounted in them may be tilted in any direction relative to one another or to the survey plane and may be turned in azimuth, optical means being provided to enable the operator to determine when the setting is such that the photographs are oriented and tilted correctly in order to solve problems 1 and 2. The bar C is raised or lowered by the foot wheel D to correspond with variations in ground height. On looking through the telescopes (above G) a black dot is seen which can be made to move in any direction on the overlap area by the appropriate movement of the wheels *E* and *F*, a movement which is reproduced by the pencil on an automatic plotting-table to the left of the machine, to which it is coupled at the desired scale by the gear-boxes on which the wheels *E* and *F* are mounted. It is therefore possible to trace the outline of topographical features by moving the dot round them on the photograph, the corresponding shapes being drawn by the pencil on the plotting-table.

The Williamson-Ross stereoscopic plotting equipment shown on Plate 8 has the successive photographs mounted in the projectors (*A*) which, by movements of the individual projectors or of the bar on which they are suspended, may be set to any position required to correspond with the tilts and azimuth of the camera at the time of photography. Variations of ground height are allowed for by raising or lowering the small round tracing-table (*B*) on to whose surface the images from two overlapping photographs are projected. The stereoscopic effect is obtained by inserting red and green colour-filters into adjacent projectors and viewing their combined images through the red and green spectacles *C*. The central point of the round tracing-table *B* is marked by an illuminated point or dot and detail is drawn by moving this mark (sliding the whole tracing-table over the main steel base table) round the objects to be mapped, a pencil vertically below the dot marking the shapes on a sheet of paper mounted on the steel table surface. The equipment at the left-hand side of the machine is an electric blower to provide air cooling for the lamp-heads of the projectors, together with a transformer and control panel to govern the illumination of the projectors.

In both types of equipment the variations of ground height are shown by verniers or micrometers recording the mechanical movements which correspond to this change of height and it is therefore possible to mark spot heights or to draw contours on the finished map. The degree of precision of such measurements depends on the characteristics of the photographs, but under good conditions it is possible to deal with differences of the order of $1:1,000$ of the flying height.

A number of other types of stereoscopic mapping equipment have been produced in the past and others are under construction, but for the present purpose it is sufficient to appreciate that all have been designed to overcome the same group of problems along roughly parallel lines, though in the future improved design and better air cameras may greatly increase the applicability and precision of these methods.

INDEX

For Product Safety Concerns and Information please contact our EU
representative GPSR@taylorandfrancis.com
Taylor & Francis Verlag GmbH, Kaufingerstraße 24, 80331 München, Germany